Praise for *The Winter Market Gardener*

Jean-Martin's passion for organic farming is contagious. He is always innovating, experimenting, and sharing his findings with others and *The Winter Market Gardener* is no exception. This book is practical, thorough, well-researched, and an invaluable resource for cold-climate growers.

— Erin Benzakein, Floret

The core challenge of growers in the north is learning how to feed our communities with fresh, local, real food in the winter. With great research and detailed charts, this book shows exactly how to do just that. I think every grower should buy it, read it, and incorporate its ideas. An indispensable book.

— Ben Hartman, author, *The Lean Micro Farm*

In this book, you will find all manner of useful planning considerations and practical tips and evaluations of structures, tools, diseases and pests, crop notes, and optimal storage conditions to help you extend your seasons and production for as long as possible through the winter months… A lovely extension to Jean-Martin's previous handbook, and a great resource for those looking to up their winter growing practices.

— Richard Perkins, author, *Regenerative Agriculture*

Drawing on the Parisienne market garden tradition and Eliot Coleman's seminal works, *The Winter Market Gardener* presents years of new research, experimentation, and practice. It covers every aspect of growing vegetables for market year round in a cold climate to build community resilience and reduce imports. It includes types of structures and covers, techniques, the tools required, and preparing products for sale. It's comprehensive, well laid out, full of information, and engaging. I particularly enjoyed reading the plant profiles—many familiar but some new ones I now have to try.

— Maddy Harland, editor, co-founder, *Permaculture Magazine*

The essential guide for winter market gardening for the new generation of growers is here!!! In the face of capricious socio-economic and environmental conditions, growing food through the winter is of utmost importance. Winter market gardening meets year-round community needs for healthy local produce and sustainable profitability for small-scale farmers.

— Zach Loeks, edible ecosystem designer, author,
The Two-wheel Tractor Handbook and *The Permaculture Market Garden*

T0286525

the winter market gardener

A SUCCESSFUL GROWER'S HANDBOOK
FOR YEAR-ROUND HARVESTS

JEAN-MARTIN FORTIER AND CATHERINE SYLVESTRE

new society
PUBLISHERS

Cover design by Diane McIntosh.
Cover image by Jessica Théroux.
Printed in Canada. First printing October, 2023.

This book is intended to be educational and informative. It is not intended to serve as a guide. The author and publisher disclaim all responsibility for any liability, loss, or risk that may be associated with the application of any of the contents of this book.

Inquiries regarding requests to reprint all or part of *The Winter Market Gardener* should be addressed to New Society Publishers at the address below. To order directly from the publishers, please call 250-247-9737 or order online at www.newsociety.com.

Any other inquiries can be directed by mail to:
New Society Publishers
P.O. Box 189, Gabriola Island, BC V0R 1X0, Canada
(250) 247-9737

| Funded by the Government of Canada | Financé par le gouvernement du Canada |

Title: The winter market gardener : a successful grower's handbook for
    year-round harvests / Jean-Martin Fortier and Catherine Sylvestre.
Names: Fortier, Jean-Martin, author. | Sylvestre, Catherine author
Description: Includes bibliographical references and index.
Identifiers: Canadiana (print) 20230452078 | Canadiana (ebook)
    20230452647 | ISBN 9780865719873 (softcover) |
    ISBN 9781550927801 (PDF)
Subjects: LCSH: Vegetable gardening—Handbooks, manuals, etc.
Classification: LCC SB320.9 .F67 2023 | DDC 635/.04b—dc23

Contents

Acknowledgments

In gratitude to all the pioneers who paved the way before us in the field of organic farming. Their contributions have been instrumental in shaping the way we practice agriculture today.

And to all the new organic growers who are joining this movement. Small-scale farming is changing the world.

Foreword

Turns out that Jean-Martin Fortier visited my farm, and I wasn't even there. In our young farming years, as we were building the operation back in the early 2000's, he was touring winter farms and stopped by to see what we were making of a hilly corner of Washington County, NY. He recently sent me some pictures of that visit and, now that I'm in Ohio, I shudder at the amount of snow we got up there in the long, cold upstate NY winters. Years later, we officially met as speakers at the Young Farmers Conference at Stone Barns. We spent a solid couple of hours at the bar geeking out over wash-and-pack building-and-flow design, and at the end, he invited me up to his new project farm, La Ferme des Quatre-Temps, to consult on the project. While most of the time was spent on systems and processes for the washing and packing side, we did spend some time on the winter production that they were doing. One of my notes in the report was about the opportunity he had to keep great records and share what he discovered.

We got started in winter growing back in 2003, when my mentor, Paul Arnold, said to me, "you know, there's this new thing called winter growing you can do." He recommended Eliot [Coleman]'s book and off we went. That first year we planted way too late, Macgyvered a heating system together, and finally got a crop in March. But we were hooked. Mesclun, spinach, radishes, and more came from that first harvest, and we were the only ones at market with those crops, resulting in our biggest-ever early-spring sales. As we learned the principles, we started breaking the barriers of what we could grow. I clearly remember being invited to a conference where the previous government extension speaker told us that growing lettuce during the winter in our zone just couldn't be done, and my presentation was about how we were doing it.

But why winter growing? Winter is cold, wet, and dark. Farmers are tired and would rather sit on a beach....

I'll be the first to admit that taking it easy isn't my strong suit, and I have always admired the way Jean-Martin and his family prioritize seasons of rest. But when it comes to winter growing, we have to really expand our perspective. This is an opportunity ripe with endless possibility, and for a growing farm, a perfect opportunity to gain traction within your market. Greens and radishes and mesclun in February. Spinach that doesn't go bad in a week for your customers at the market, plus a consistent growing season to supply produce to your CSA members. And there's really nothing like coming into a tunnel after a –10°F night and seeing all the plants slowly waking back up. The air is humid and smells like soil. Even though the wind may buffet the outside, inside is peaceful and full of hope. It's also worth noting that the idea of winter growing and using winter as a season of rest are not mutually exclusive. It's going to look different for every farm, and every farmer. For our farm, it's a short work week and open market, a skeleton crew we can rely on when we need to head out for family gatherings or a sandy moment in the sun. But it's mostly a time when we work a few hours in the tunnels harvesting and covering/uncovering, and then reading up on ideas to implement next year, or being honest with ourselves about what just didn't work for our team the previous year. It's reflection and rest, but also movement. But the movements are slower, because as Jean-Martin and Catherine Sylvestre—La Ferme des Quatre-Temps' farm manager—are about to lay out, slower is better. It's better for the plant, the product, and the producer. This isn't peak season of hitting your pillow at night bone-tired and a little dehydrated. No; winter is a saunter. And after the spring, summer, and fall, we are happy to throttle down, and yet still have a source of revenue, a fresh product to delight our customers, and, frankly, some soil and life to work with. Most of us didn't come into farming because we wanted to take it easy. We wanted an honest life of work that mattered. And winter growing? It really matters.

What I would give to go back in time with this book to that first season of winter growing. The charts, illustrations, and graphs clearly lay out the principles, systems, and processes to be a successful winter grower. Jean-Martin and Catherine did end up keeping great records, and that collection of research will help all of us take our winter growing to another level. But most importantly, this book was written by two people who have devoted their lives to helping feed communities better. Winter is no longer off-limits, it's open to all of us.

— Michael Kilpatrick, Educator at Growing Farmers,
Farmer at the Farm on Central

Introduction: Yes, We Can Grow Vegetables Year-Round...and Why We Should

In Quebec and most northern parts of the US and Canada, comes winter. This seasonal change is inevitable and leads us to easily abandon the idea of eating in-season year-round. In most people's minds, the fresh season starts in June and lasts until October, perhaps a bit earlier the further down the longitude line you live. "See you next year! We're going to miss your tasty veggies!" is what most growers hear every year when outdoor markets come to an end. Come fall, as colder nights start kicking in, everyone suddenly stops celebrating local farming. The weekly act of buying fresh produce from a neighboring farm suddenly comes to a halt. This is the sad reality of growing vegetables in northern climates or so it was...

In many small-scale farms, the status quo is evolving. When the season stops, we just keep planting and harvesting. This was the famous call to action of the pioneer of winter farming, Maine's own Eliot Coleman. A call that has been answered by thousands of organic growers in all parts of the northern states and provinces who benefit from a strong market for local products. This is where our story starts.

The "eat local" movement has made some tremendous strides in the last twenty years, creating in many communities a market for year-round local products.

For the last ten years, I've been trying to figure out how to grow year-round and out of peak season, and doing so with a profitable, low-tech approach. My journey in winter farming first started by visiting growers in Europe and parts of the US and Canada and by attending every farmer conference I could. Along the way, I took notes about the different approaches and results that growers would gladly share.

Then my professional life took an interesting twist. After ten years of successfully running my own market garden operation, I was asked by a Canadian businessman and philanthropist to develop a farm like no other, one that could demonstrate an alternative way of producing food in a holistic and ecological manner. This farm—Ferme des Quatre-Temps—would provide nourishing products for its local clients and become a training ground for young Canadians seeking to start their own farming projects. It was here that the idea of developing techniques for year-round farming in northern climates blossomed. And so, since 2016, I've been testing and trialing different patterns in order to achieve what became the farming manual you now hold in your hands. Along the way, a former apprentice of Ferme des Quatre-Temps and professional agronomist, Catherine Sylvestre, took my place as the farm's main vegetable grower, and together we conducted research on best practices for winter farming. Catherine and I are what you can call... geeks. We love winter farming because of the challenge, and for many years, the advancement of our techniques was driven by this quest to learn and improve our practices. This was true until the pandemic hit and our research at Ferme des Quatre-Temps came under the light of a new paradigm. What if winter farming became essential?

Going local year-round

Throughout the year, our grocery stores receive an abundance of fruits and vegetables from around the world, delivered via efficient and relatively cheap transportation systems. Today's supermarkets know no seasons, and when environmentalists criticize the ecological costs of these systems, their complaints are heard as such...complaints! The modern lifestyle allows us to eat however we want, whenever we want, and we are no longer condemned to the fate of our ancestors, who ate nothing but potatoes, carrots, and turnips in the winter months.

This narrative was perhaps true before the effects of the COVID-19 pandemic awoke many folks to the fragility of our modern grocery store supply chain. We then learned that in the Northeast, roughly 70 percent of vegetables sold in grocery stores come from California, Mexico, or elsewhere in the world. When this supply chain was compromised, we also learned that it takes about three days to deplete the supply of fresh fruits and vegetables. Suddenly, we could see that these imported products came with some disadvantages. Perhaps the vulnerability of this food system was worth planning against.

Although the consequences of the pandemic and the temporary shutdown of borders did not reach red alert with regards to food accessibility, they served as a warning, an eye-opener for some policy makers who decided to take action. In Quebec, one of the main policies was a massive investment program to double the number of greenhouses within five years. Suddenly it became a matter of national security to grow in winter. Unfortunately, the idea only got picked up by large-scale producers who already could invest in building new greenhouse complexes. These producers grow summer crops in monoculture regardless of the season.

Catherine and I decided then to propose our alternative: Is it better to invest $20 million to bolster a twenty-hectare tomato greenhouse complex or, instead, to invest the same amount towards better equipping and educating fifty family farmers, so that they can use greenhouses and extend their growing season to provide a diversity of seasonal and local produce?

Why not aim for real food sovereignty and decentralize the production with thousands of smaller greenhouses all over the province of Quebec?

This would, we imagine, create the independence we seek for our province. And so we decided to share our insights and, most importantly, our production techniques. The idea, as with many new things, is to first convince by showing the evidence.

Winter growing, the kind we practice at Ferme des Quatre-Temps, requires little energy and is in tune with the seasons. Many vegetables can withstand cold and even below-freezing temperatures when protected from icy winds. These are the plants we work with, and the goal is to master their growth in new, more challenging conditions. We have learned to select cold-hardy and disease-resistant cultivars (varieties), to protect our crops with simple and affordable shelters, to plan our successions properly, and to adjust crop densities to account for inhibited plant growth and diminished sunlight. From spinach, to celery, to parsley, we grow nearly thirty different vegetables that can be harvested in winter months. Our experiments with heating at different levels also gave us some insights into growing these veggies economically. The factor that contributes to the success of our winter farming depends on one variable: the empirical knowledge of humans who are convinced that we can grow vegetables differently.

If you are reading this, you are likely a vegetable grower or a consumer hoping to reinvent northern agriculture as we know it. We invite you to follow us and join in this movement that is sure to change the agricultural landscape. Eating according to the seasons is both beautiful and wholly logical. We never cease to be amazed, from the first sweet spring radishes to delicious heirloom tomatoes in the summer, flavorful squash in the fall, and spectacular spinach in the winter.

Without further ado, let's dive into *how* we can get there.

3

— JM Fortier

Ferme des Quatre-Temps, or Four Season Farm, refers to a Canadian native flower with four petals, also known as bunchberry. Each petal represents one of the four very distinct seasons in Quebec where we farm. The name is also an homage to Eliot Coleman's famous Four Season Farm in Maine, with his blessing.

The concept behind Ferme des Quatre-Temps, located in Hemmingford, Quebec, was developed by Jean-Martin Fortier and a team of permaculture designers. Their objective was to strike a sweet spot between intensive production of organic farming while creating a landscape designed for biodiversity. The farm is holistic as it is thought of as a whole that includes gardens, pastures for mob grazed beef, free-range chickens, and hogs that have access to rotational grazing. The site includes nearly ten acres of land used for biointensive vegetable production, as well as greenhouses and high tunnels for winter productions.

Every year, Ferme des Quatre-Temps welcomes apprentice market gardeners who spend two seasons learning the techniques and methods developed on the farm. A Quebec TV series called *Les fermiers* features this unique educational program that gives each participant the opportunity to take on a management role and oversee all operations.

The knowledge developed on the farm is also advanced and shared through the Market Gardener Institute. The goal of the Institute is to equip and support the new organic growers of today and tomorrow who are changing the world by growing healthy food for their communities. The Institute disseminates this knowledge through various media, including its online course, The Market Gardener Masterclass.

The Knowledge
and Skills of
Northern Growers

❄

The first salads or radishes are harvested in January, the first carrots or turnips are harvested in March, and the first ripe melons are in April.

— Joseph Vercier, *Culture potagère*, 1911

Some ideas that may seem avant-garde are in fact quite old; this is certainly the case with northern agriculture. In 1942, Abbé Maurice Proulx produced a short film showing how winter vegetables could be grown using hotbeds (*couches chaudes*).

As early as January, growers would dig squares roughly three feet deep into the snow, then stake twelve-inch boards into the ground. Into this hole, they added an eight-inch layer of hot manure and covered it with topsoil. The soil would be warm, having been stored in bags left indoors since the fall. Young tomato, onion, and lettuce seedlings were then planted in the soil and kept warm by the manure. The hole was sealed using a frame with glass panels (like a window) built for this purpose. On sunny days, the frames were propped open to prevent overheating. This simple shelter, combined with the heat generated by manure, allowed growers to harvest crops several months before other gardeners. The tomatoes were later transplanted into the vegetable garden, while lettuce and onions could be eaten in March and April.

Our ancestors knew how to extend the harvest season using nothing but simple shelters. They also had an abundant supply of fresh manure, which is an indispensable heat source for hotbed systems.

In France, Abbé Proulx studied the methods of Parisian market gardeners who had mastered the art of growing vegetables in the off-season. For more than 200 years, these market gardeners fed the millions of people inhabiting the French capital with vegetables and several types of fruit grown year-round. On small diversified farms that harnessed knowledge and skills to grow crops in an entirely ecological manner, a single plot could see eight successions in one year. And at this time, petroleum-based fertilizers, pesticides, and herbicides were not yet available!

Yet even in France, this approach gradually fell into oblivion. As it became more efficient to ship perishable goods, produce could quickly be moved from one region to another and, eventually, from one continent to another. Local off-season agriculture was replaced by imported fruits and vegetables grown in warmer climates and parachuted into major cities like Paris.

Fortunately, the extraordinary know-how developed by 19th-century French gardeners was not entirely forgotten. American Eliot Coleman is without a doubt one of the pioneers who revitalized these methods. A contemporary thought leader for many vegetable growers in the organic farming world, Coleman spent more than forty years on his farm in Maine modernizing methods for market gardening. He documented his observations in several influential books like *The Winter Harvest Handbook*, which remains the best reference for growing vegetables in the cold season.

Obviously, Coleman and Abbé Proulx are not the only ones who applied and documented these methods. With a little research, you will discover a trove of resources. Today, the revival and dissemination of this expertise have become even more relevant as the agricultural industry searches for a greener path forward. The book you are holding in your hands and our experiments at Ferme des Quatre-Temps are a part of this process.

Abbé Proulx was a priest, agronomist, and filmmaker who documented the transition from peasant farming to industrial agriculture for more than thirty years. As a scholar, he traveled extensively to learn about European methods for growing vegetables and was the first to make educational short films about good agricultural practices. His documentaries are uniquely beautiful and portray a not-so-distant past.

Photo courtesy of the Archives de la Côte-du-Sud et du Collège de Sainte-Anne.

Eliot Coleman is one of the pioneers of the American organic movement and lives in Maine, just a few hours from the border with Quebec. Thousands of American market gardeners, especially in the northeastern United States, used the tenets of his work to develop new strategies and run their own farms. The result has been a lively and captivating proliferation of disciples who have shared their discoveries (and their mistakes!) over the past twenty years.

Coleman largely draws from the expertise of 19th-century Parisian gardeners and their approach to small-scale intensive vegetable production. In recent years, he published new techniques and methods for winter growing. These publications marked a critical milestone; they democratized winter growing, allowing the public to access previously unavailable information and furthering the northern market gardening movement as a whole.

Photograph by Barbara Damrosch. Photo courtesy of Eliot Coleman.

Sharing Our Collective Knowledge with New Farmers

Founded in 2015, Ferme des Quatre-Temps is an educational farm that aims to train future growers according to a regenerative market gardening business model. The farm has two sites: one in Hemmingford, southern Quebec, and the other in Port-au-Persil, in the northern Charlevoix region. At both locations, experiments are strongly encouraged and celebrated.

Since the farm's inception, winter growing has been an integral part of the methods we actively strive to adopt and improve. Through this approach, we aim to find ways to make the least productive months profitable, establishing best practices for successful winter growing in Quebec.

We went through many rounds of trial and error, some resulting in frozen vegetables and freezing hands, and others revealing surprising successes. Eventually, we implemented procedures that, in recent winters, have shown proven results. In this book, we will shed light on our past efforts, and we hope that it will inspire many people to consider adopting these methods in their own market gardens.

Through our experiments, we identified five key principles that guide our approach to winter growing. We will further explore these concepts throughout the chapters of this book.

- **The most significant limiting factor for winter vegetables is a lack of sunlight, and not a lack of heat.**

- **Many vegetables and cultivars can withstand freezing temperatures.**

- **It is possible to increase the cold hardiness of some crops.**

- **By using simple shelters and layering them like onion peels, we can create favorable microclimates that allow vegetables to grow in winter.**

- **By working with minimal heating, we can limit the investments required to make winter crops profitable.**

Before going any further, we want to acknowledge the incredible contributions made by an entire community. They enhanced our collective knowledge and skills, helping us to uncover these essential concepts. Our approach to farming and this book would not have been possible without the long-standing knowledge transfer network established by farms in the United States, Ontario, and Quebec, and their combined expertise. We wish to thank them for their generosity in enhancing our collective understanding and driving our movement forward.

It is estimated that there may have been up to 10,000 small farms surrounding 19th-century Paris. These gardens managed to feed the city without modern technology like electricity, heat from fossil fuels, and plastic. This shows us that it is possible to transition to a form of agriculture that is different from the one we know.

Of course, the lives of these vegetable growers were certainly a far cry from our modern-day context. Some worked up to 18 hours a day! But by combining simple shelters, a strong understanding of plants, adaptability, radiant heat sources (manure, at the time), and incentives to produce (business was good in Les Halles), these growers developed an approach to production that remains a useful reference even today. And it is not so far removed from what we currently know.

We are careful here to avoid the trap of nostalgia, claiming that everything was perfect in the past. But we can reliably say that these vegetable growers were one step ahead of their modern-day counterparts who operate only six or seven months a year. The general lack

Photo courtesy of the Bobigny commune archives, in France.

of local winter production has left a void and enabled large-scale commercial and global operations to fill this market void with imported produce requiring a lengthy, fragile, and fossil-fuel-intensive supply chain.

Most books that explore methods used in the past are only available in French, the language of the Parisian market gardeners. However, we've found an English equivalent written by John Weathers in 1909 called *French Market Gardening*. Weathers recognized that French gardeners were one step ahead of British gardeners in terms of intensive market gardening techniques and season extension.

To bridge that gap and increase food sovereignty in English cities such as London, he took upon himself to document and publish the French gardeners' techniques in English. The level of detail in his book is quite advanced; we can see how inventive and resourceful the Parisian gardeners were, and we can learn from them even now and actualize their techniques!

Using Simple Shelters to Protect Crops

❄

Necessity is the mother of invention.

— Plato

Growing and harvesting vegetables in the heart of winter is an adventure that starts with protecting crops from bad weather, especially from cold winds. It's crucial to understand how the wind, through its drying effect, can damage them to the point of threatening their survival. Like us, vegetables can be comfortable in the cold, but we run into trouble when conditions are exacerbated by windchill. Drastic temperature fluctuations can also make crop survival precarious. The key to success is creating microclimates that protect crops from harsh winter weather.

As explained previously, 19th- and early 20th-century vegetable growers used cold frames to protect their crops from winter airflow. Before that, vegetable growers in Paris used glass bells (*cloches*). These protected each tomato plant, each lettuce head, or any other vegetable they tried to force into producing earlier in the season, while conditions were unfavorable for growing.

The cold frame method can still be useful today, and it is not uncommon to meet older gardeners who rely on it. Glass bells, however, are now a thing of the past. In our day and age, who would bother to remove and put back each bell in their garden?

With advances in technology, tunnels have now replaced cold frames, and the bells were replaced by what are known as row covers. Today, the combined use of these two simple technologies is at the heart of winter growing for people who farm in sync with the seasons.

The truth is, it's easy to grow vegetables in the winter in a heated greenhouse. With the help of ever-evolving technology, we can get around seasonal realities. A well-insulated modern greenhouse, equipped with a heating system and grow lights, can recreate the perfect climate, no matter how harsh the weather is. The only constraint then becomes the cost associated with the initial install and climate regulation. But the cold nights of our northern winters make it extremely expensive to recreate optimal growing conditions.

This explains why, in Quebec, vegetables grown year-round are mainly tomatoes, cucumbers, peppers, and some climbers like beans. These products yield over quite a long period (six months or more) and generate the highest revenue per square meter of crop surface. The greenhouse industry has developed highly precise crop management sequences to optimize growing conditions for these vegetables and ensure that production is profitable.

That being said, it is also *extremely* energy-intensive to grow these heat-loving vegetables in winter. It requires modern facilities that are highly expensive to deploy. As a result, industrial greenhouse complexes tend to be the only operations capable of taking on this kind of project.

In recent years, several investments have been made to increase the production of these crops grown for the Quebec market and even for export. In fact, in 2021, to promote Quebec's food autonomy, the provincial government granted a $30-million loan to a large greenhouse company to build the biggest greenhouse complex in Quebec, with a 37-acre (15-hectare) footprint. Without disparaging the initiative, take a moment to imagine how the landscape would look if these funds had financed the development of 200 small vegetable farms. They could easily add three to four months to their production season by acquiring small greenhouses. And for the rural communities that these small farms serve, the economic impact would be overwhelmingly positive.

We firmly believe that in a northern climate, a local, diversified, and quality vegetable offering doesn't have to come from mass production in heated greenhouses. This means choosing cold-hardy crops that don't require artificial light. With this strategy, small farms can capitalize on the protection provided by a greenhouse or a simple shelter in winter to grow vegetables year-round, with minimal or no heating. Vegetables that are well-suited to colder temperatures include spinach, kale, mesclun, bok choy, arugula, and Swiss chard.

Simple Shelters

At Ferme des Quatre-Temps, we have perfected the art of using the right simple shelter at the right time. We install various equipment, such as caterpillar tunnels, low tunnels, permanent high tunnels, and row covers, depending on the needs of the crops and the time of the season. Each shelter is suited to a specific use, and we sometimes combine them. The easiest way to understand their purpose is to explain the role of each one.

Crop Yields in a Three-Season Greenhouse
(April to October)

Crops	Yield per bed	Revenue $/bed	Revenue $/ft.² ($/m²)
Cucumber	1,600 units	$2,400*	$9/ft.² ($100/m²)
Eggplant	1,000 lb. (454 kg)	$3,000	$12/ft.² ($130/m²)
Pepper	450 lb. (204 kg)	$2,700	$11/ft.² ($115/m²)
Pole bean	150 lb. (68 kg)	$1,500	$6/ft.² ($65/m²)
Tomato	2,000 lb. (907 kg)	$4,100*	$16/ft.² ($175/m²)

* These crops cover two beds.

Crop Yields in an Unheated Greenhouse
(November to April)

Crop	Yield per Bed	Revenue $/bed	Revenue $/ft.² ($/m²)
Arugula	60 lb. (27 kg)	$720	$2.80/ft.² ($30/m²)
Bok choy	300 units	$900	$3.70/ft.² ($40/m²)
Kale and Swiss chard (bunched)	300 bunches	$1,050	$4.20/ft.² ($45/m²)
Mesclun (Salanova)	60 lb. (27 kg)	$720	$2.80/ft.² ($30/m²)
Spinach	60 lb. (27 kg)	$720	$2.80/ft.² ($30/m²)

▲ This is the revenue we generate with greenhouse crops grown in summer, compared to our winter production in greenhouses with little or no heating.

Revenue per square foot (sq. m.) is lower for vegetables grown in a greenhouse with little or no heating, but this method requires a much smaller investment, both in terms of infrastructure and heating costs.

◄ Swiss chard, kale, Asian greens, and mesclun are some of the vegetables suited to colder temperatures. They grow well despite a low energy supply. With these fresh greens, we can improve our vegetable offering in winter months.

Most vegetables grown in modern greenhouses in the winter, like tomato, cucumber, and egg-plant, require artificial lighting, which is expensive and generates significant light pollution. For the towns and ecosystems surrounding the complexes, there are negative repercussions; they emit light emissions for up to twenty hours a day, and disrupt circadian rhythms in both animals and humans. While the horticultural industry is increasingly relying on LED lights, the long-term health effects of consuming vegetables produced with artificial lighting are not yet known.

Photograph by André Muller, iStock Photo.

ROW COVERS

Row covers are large pieces of nonwoven polymer fabric that help when ambient temperatures are nearing 28°F and 27°F (−2°C and −3°C). Made of permeable material, they allow air, light, and water to pass through. When protecting crops, row covers retain heat near the ground, which increases temperatures while helping to conserve moisture. In doing so, they provide a few additional degrees of frost protection. By acting as physical barriers, row covers also help protect seedlings from weather events, like driving rain and strong winds, and from pests.

Row covers are available in different thicknesses (calculated in ounces per square yard or grams per square meter). Your choice of product will depend on the situation and season. On our farm, both in spring and fall, we use row covers with a 0.55 oz./sq. yd. gauge (19 g/sq. m). They are durable (when handled with care, they will last more than one season) and provide 85 percent light transmittance compared to an uncovered crop. This cover, called P19, also provides an increase in temperature of approximately 3°F to 4°F (1.5°C to 2°C).

In the event of a severe frost, the row cover protects vegetables from the cold. By combining several tunnels and covers, we can gain a few extra degrees, which is the difference between a crop being killed or surviving a night of below-freezing temperatures. High tunnels provide wind protection and create a greenhouse effect that warms the thermal mass of the soil during the day. At the end of the day, when the hot air rises, the row cover traps that accumulated warm air, keeping it close to the crops.

For optimal results, row covers are spread out over hoops made of galvanized steel wire that prevents them from touching the crop's leaves, which could cause frost damage. In winter, we use row covers inside other shelters for additional protection against the cold. Sometimes, we even use two or three layers of row cover.

Good row cover management is essential in unheated greenhouses and tunnels as it helps limit overnight heat loss. We recommend using the following guidelines to ensure success:

When the temperature in the greenhouse is near freezing (32°F, 0°C):
Set up one P19 row cover on the crops.

When the temperature in the greenhouse is 23°F (–5°C):
Set up two P19 row covers on crops, layered one over the other.

When the temperature in the greenhouse is 14°F (–10°C):
Set up three layers of P19 row cover on crops.

When day length is less than 10 hours: Leave the row covers on the crops at all times. Heat built up under row covers will contribute to a significant increase in overnight temperatures around the plants. During the day, the sides of row covers can be raised slightly to promote ventilation.

When day length is longer than 10 hours: Remove the row covers between 9 a.m. and 3 p.m. on sunny days, unless temperatures in the greenhouse are at or below freezing. At this time of year, daylight hours are long enough to justify removing row covers.

LOW TUNNELS

In the fall, low tunnels are the perfect low-cost solution for extending the season until the first big snowfall. They cover a smaller surface area than caterpillar tunnels (two beds instead of four), but are stronger, easy to install, and highly affordable. These simple structures can be moved throughout the farm and can withstand some snow loading.

Low tunnels are made from galvanized steel tubing (EMT), which is sold in hardware stores and shaped using a pipe bending machine. Compared to hoops that hold up row covers, low tunnel structures are much stronger, and they can therefore support a polyethylene film (transparent greenhouse plastic).

To install the tunnel, we start by placing rebar into the ground. Then, we fit the hoops over the rebar and cover the frame with a layer of transparent plastic. The tunnel is secured by ropes that are tied to the bases of various arches and that run over the plastic. At each end of the tunnel, the plastic extending past the hoops is roped to a piece of rebar.

The advantage with low tunnels is that they create a greenhouse effect, are more affordable than high tunnels and greenhouses, and can withstand some snow loading. Still, this remains only a three-season shelter.

CATERPILLAR TUNNELS

Unlike low tunnels, caterpillar tunnels cannot support any snow load. Every year, we have to remove their plastic covering before the first big snowfall. This disadvantage doesn't deter us from using them, however, as they are one of the simplest, most affordable, and effective solutions to extend the growing season.

When compared to high tunnels, the biggest advantage of caterpillar tunnels, aside from their price tag, is their mobile design. Because these structures can easily be assembled and disassembled, they can be moved anywhere in the garden at any time of the season. In early spring, we position them over our first seedings in the field, to protect them from those final cold nights. Then we move them onto tomatoes, peppers, and melons, which always need extra heat in the summer. When the first frost arrives, we move them once again to protect fall crops that have begun growing in the field. Vegetables like spinach, mustard, and baby kale can withstand several frost events and continue to grow thanks to the microclimate created by a caterpillar tunnel.

Caterpillar tunnels are one of the simplest and most effective solutions for extending the growing season.

To build a caterpillar tunnel, you can choose between many techniques and materials. The simplest approach requires only galvanized steel tubing (bent using a pipe bending tool), rebar, straps, rope, and polyethylene film. The arches are installed 5 feet (1.5 m) apart and are anchored onto the rebar, which is sunk halfway into the ground. To reinforce the structure, a strap connects every arch at its apex. At each end of the tunnel, that strap is tied to two 6.5-foot (2 m) T-posts that are securely planted into the ground at a 45-degree angle. The plastic then goes between these two bars, which are tied together to maintain tension. Lastly, like with the low tunnel, ropes go over the transparent plastic and are tied to the bases of the arches. Using a tensioning knot in each rope, the plastic can then be pulled taut. The tunnel looks like a long caterpillar, which is how the structure got its name.

For ventilation, we open the tunnel by manually lifting the plastic along the sides, and the ropes keep the plastic from moving.

HIGH TUNNELS

A high tunnel (also known as a passive solar greenhouse, hoop house, or polytunnel) is a permanent structure made from semicircular steel arches, which are bolted and covered with a polyethylene film. Unlike greenhouses, high tunnels are not tall and have a simple, often more rounded structure. Generally, they are 20 to 26 feet (6 to 8 m) wide, and the apex is no more than 10 feet (3 m) high, while length is variable, ranging from 50 to 150 feet (15 to 45 m). At Ferme des Quatre-Temps, we've standardized all our permanent tunnels to a 100-foot (30 m) length.

These tunnels are sometimes referred to as cold tunnels because, unlike greenhouses, they are unheated and are therefore less insulated. Of course, this means they are also more affordable. The tunnels have roll-up side openings, and large doors at the front and back, which are more than enough to provide good ventilation in the tunnel.

Unlike low tunnels and caterpillar tunnels, high tunnels cannot be moved because of their more robust anchoring system. When properly installed, they can withstand the weight of a heavy snowfall without collapsing. This makes them excellent four-season tunnels. Their versatility throughout the growing season is an undeniable asset, as they protect early crops in spring and late crops at the end of the season, and also provide the perfect summer conditions for heat-loving crops. Plus, when fall comes along, you can plant winter vegetables in the high tunnel, to provide continuous production until the following spring.

Permanent tunnels come in many shapes and sizes, and most greenhouse manufacturers sell some kind of high tunnel. The price for a new model depends on the type of reinforcements, as well as the width and length; in any case, this tunnel is one of the best investments a farm can make. Revenue generated by crops grown this way will surpass construction costs within a few seasons, if not the first one.

Because these tunnels are permanent structures, you need to carefully consider their location, especially with regard to soil drainage. In the spring, poor drainage will cause beds to stay wet for too long and delay early crop plantings. The best solution is to install agricultural drains along the perimeter of the structure. Leveling the ground is also effective but requires heavy soil work. Furthermore, do not install tunnels near other buildings or large trees, as they could block sunlight part of the day. When deciding where to place a tunnel, consider an orientation that maximizes sun exposure and that is parallel to prevailing winds, to minimize snow accumulation on the shelter.

Permanent tunnels are simple and affordable shelters that provide profitability in winter and do not require excessive heating or complex technological facilities.

High Tunnel *vs* Greenhouse

In both cases, we are referring to permanent structures that protect crops from cold winds and create a favorable climate for growing. If you're new to this, it's easy to get confused. Unlike high tunnels, greenhouses are heated, and are therefore better insulated. They can be made of glass or two layers of plastic, which are constantly kept apart by a blower that pumps air between them. Greenhouses also have several systems that allow for precise climate control, and they can be set up with equipment that automates tasks.

The length, width, and height of a greenhouse will vary between models. For most small organic farms, this structure will be no larger than 35 feet × 100 feet (10 m × 30 m).

COMPARING DIFFERENT SHELTERS

Combining shelters helps to increase a crop's capacity to withstand the cold. This table shows the outside temperature at which bok choy, spinach, lettuce, and arugula will die, depending on the level of protection provided.

Cold-hardy crops	Outside Temperature—Killing Frost, in °F (°C)		
	Field (no protection)	High tunnel	High tunnel + P19
Arugula	21°F (−6°C)	7°F (−14°C)	3°F (−16°C)
Bok choy	25°F (−4°C)	10°F (−12°C)	3°F (−16°C)
Lettuce (Salanova)	21°F (−6°C)	7°F (−14°C)	0°F (−18°C)
Spinach	11°F (−12°C)	−4°F (−20°C)	−11°F (−24°C)

Shelters available on the market provide different levels of protection against the cold and the elements. Those listed in the following table reduce temperature differentials by creating a more stable climate, which improves a crop's chances of survival. When purchasing a shelter, however, you'll generally have to pay more per square foot if you want to ensure a greater heat gain.

Thermal gains, in degrees Fahrenheit and Celsius, depend on outdoor factors like wind, sunlight, ambient temperature, and humidity. As a result, we cannot guarantee an exact heat gain; the data presented are used to provide a general idea.

Shelter	Cost per ft² (m²)	Average thermal gain in °F (°C)	Features and uses
Row cover	± $0.30 (± $3)	Gained: 2.7 to 3.6°F (1.5 to 2°C)	• Can be used in all contexts (as a standalone or inside another shelter) • Can be layered to increase its effect (note that combining several layers will block sunlight, so it's important to remove the covers on sunny days) • Contact with crop foliage should be kept to a minimum during frost events • Is permeable: does not offer full protection against the weather (snow, wind, etc.), unlike the other structures • Highly mobile shelter
Low tunnel	± $1.10 (± $12)	Gained: 5.4 to 7.2°F (3 to 4°C)	• Covers 2 beds: 30 inches (75 cm) wide, with a 12 inch (30 cm) aisle • Easy and quick to install • Easy to move • A very inexpensive structure per sq. ft. (sq. m) • Weatherproof • Can even be used as a second shelter within a greenhouse, in winter
Caterpillar tunnel	± $1.60 (± $17)	Gained: 9 to 13°F (5 to 7°C)	• Covers 4 beds that are 30 inches (75 cm) wide • Easy to move • Can accommodate a suspended irrigation system • Is perfect for start-up farms • Excellent structure for extending the season at an affordable price • Tolerates very little snow load and must be cleared or disassembled before heavy precipitation • Is not recommended for sites affected by strong winds

Shelter	Cost per ft² (m²)	Average thermal gain in °F (°C)	Features and uses
High tunnel	± $2.00 (± $22)	Gained: 11 to 16°F (6 to 9°C)	• Covers 5 beds that are 30 inches (75 cm) wide • Permanent structure without a heating system • Possible to add automated systems like motorized roll-up sides controlled by a thermostat • Stays anchored even in strong winds • Tolerates some snow, but must be cleared in the event of heavy precipitation
Heated greenhouse	± $9.30 (± $100)	Significant heat gain thanks to a heating system (temperature regulated by a climate control system)	• Covers 8 to 11 beds, each 30 inches (75 cm) wide • Permanent structure with a heating system • A greenhouse climate control system can be added to automate heating and ventilation • Good energy efficiency, due to good insulation and a considerable air mass • Can withstand snowfall

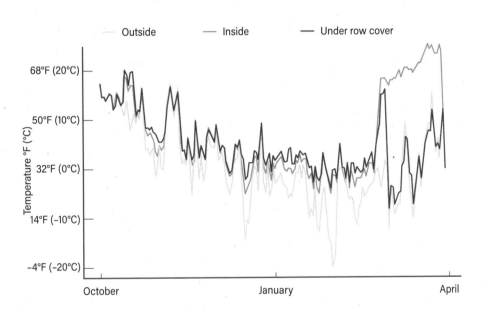

Comparing temperatures: outside and inside an unheated greenhouse, as well as under a row cover placed over the crops

This graph shows the temperatures recorded at three different locations in the winter of 2019–2020 at Ferme des Quatre-Temps. The pale blue line shows the temperature recorded outside an unheated greenhouse in the months of October to April. The gray line shows the temperature recorded inside this unheated greenhouse in the same months, and the green line shows the temperature recorded under the row cover placed over the crops in the unheated greenhouse.

We can see that the greenhouse provides an increase of a few degrees when compared to outside temperatures. The same applies to the row cover protecting crops. Combining several shelters inside an unheated greenhouse is an inexpensive strategy that often makes the difference between a crop being killed or surviving a night of below-freezing temperatures.

The Challenge of Winter Heating Costs

The economic viability of a winter growing season depends on low production costs, which indirectly means investing in simple technology. This explains why, for the past thirty years, farms that have embarked on the adventure that is winter growing have opted to leave high tunnels unheated. Even though the cost of organic vegetables sold locally has been rising, so is the cost of all fuels used in greenhouse production. This equation leaves us with the obligation to find low-tech solutions for crop protection and climate enhancement.

These will include, for instance, using row covers or even a smaller greenhouse in high tunnels, and going for double layers for better frost protection. The other common strategy is adding more tunnels on the farm. Both are proven strategies that provide growers with the opportunity to gradually expand their offering as they evolve. Each season, they can put up one or two new tunnels, according to the operation's financial capacity.

But what happens if we heat the crops instead of just protecting them: Is it profitable? According to our experiments at Ferme Quatre-Temps, the answer is yes and no…

When heating a greenhouse, insulation with a double layer of plastic and a blower is a must, as is perimeter insulation. Energy efficiency is the number one rule of action to diminish fuel costs, regardless of the type of heater or fuel. At Ferme Quatre-Temps, our greenhouses are well-insulated, and equipped heaters have perforated diffusion tubes. This system is common in conventional tomato operations and makes it easier to circulate warm air. Our greenhouses are also quite tall, thus holding a large air mass. As a result, they provide near-optimal conditions for the use of a conventional unit heater.

SCENARIO 1: GREENHOUSE HEATED TO 54°F (12°C)

Cold-hardy plants grow best at 54°F (12°C). With a classic propane heating system, the costs to maintain this temperature in our 30-foot-by-100-foot (9 m × 30 m) greenhouse averaged ±$400 per week. We harvested these three crops three times that winter, with a total yield of roughly 660 pounds (300 kg) per greenhouse. In this scenario, 71 percent of the revenue went towards paying to heat the greenhouse. The sale of greens, even at an optimal price point, couldn't justify the propane costs to heat our greenhouse all winter.

SCENARIO 2: UNHEATED GREENHOUSE

In a second greenhouse, where no heat was introduced, total crop yields were approximately 265 pounds (120 kg), equivalent to 40 percent of the yields from our heated greenhouse. It was obvious that when heating is entirely eliminated, the decrease in production is drastic. The overall quality is also lowered. However, and this is very important, this scenario proved to be more profitable due to the lack of heating costs. Workers are required for handling row covers day and night to maximize their effectiveness, which adds labor costs. These need to be factored in when assessing profitability.

SCENARIO 3: MINIMALLY HEATED GREENHOUSE, 37–41°F (3–5°C)

In a third greenhouse, we provided only minimal heating to keep temperatures just above freezing. In this scenario, our overall yields were roughly two-thirds of the yield from our fully heated greenhouse. Quality was also overall better than in the unheated greenhouse. The results here provided a good balance: ample yields at a reasonable cost, with 40 percent of revenue going towards heating costs and a roughly 45 percent increase in production compared to the unheated scenario.

In the winter growing experiments in the unheated greenhouse, we lay a row cover over the vegetables and run perforated tubes down the aisles under this cover to maintain airflow while heating only a restricted space. With this method, more heat stays near the base of the plants. The sensor collecting data for the climate controller is installed under the cover, to record the temperatures that the plants are experiencing.

Scenarios and Costs/Revenue

	Scenario 1 Greenhouse heated to 54°F (12°C)	Scenario 2 Unheated greenhouse	Scenario 3 Greenhouse with minimal heating 37–41°F (3–5°C)
Gross revenue	660 lb. (300 kg) at $13.65/lb. ($30/kg) = $9,000/greenhouse	265 lb. (120 kg) at $13.65/lb. ($30/kg) = $3,600/greenhouse	440 lb. (200 kg) at $13.65/lb. ($30/kg) = $6,000/greenhouse
Heating costs	$400 propane/week × 16 weeks = $6,400 per winter	$0 propane/week × 16 weeks = $0 per winter	$150 propane/week × 16 weeks = $2,400 per winter
Ratio of heating costs to revenue	(6,400/9,000) × 100 = 71%	(0/3,600) × 100 = 0%	(2,400/6,000) × 100 = 40%
Net revenue	$2,600	$3,600 (Minus additional labor costs)	$3,600

Comparing Three Greenhouse Heating Strategies

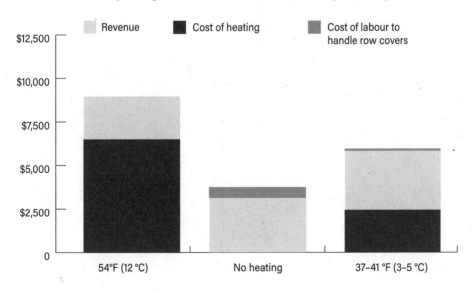

Although we were quite surprised with the results of our experiment, the conclusion is enlightening. In addition to generating good yields from cold-hardy crops, a strategy of minimally heating a greenhouse in the winter added undeniable benefits. First, it increases efficiency in the workforce by reducing time spent handling row covers day and night, but most importantly, supplemental heating protects crops from damage due to extreme temperature variations. This represents a critical safety net: no crop will be killed by an overnight freeze. When compared to the unheated approach, the minimal heating scenario also makes it possible to grow a greater variety of vegetables. This solution allows growers to experiment with cold-sensitive crops without having to suffer losses caused by a significant frost event.

SWEET SPOT

As a result of our experiments, we have concluded that heating a greenhouse to 37°F to 41°F (3°C to 5°C) creates the ideal conditions for a winter production that is both worthwhile and enjoyable and, therefore, have invested our resources in this method. Winter production may seem to generate only modest revenues, but remember that growing greens requires very little work, and that these crops will allow a grower to gradually sell off vegetables harvested in the fall (i.e., by combining storage crops with greens and fresh vegetables harvested in the greenhouses). Plus, using a combination of several shelters (row covers, caterpillar tunnels, high tunnels, low tunnels, and greenhouses) can increase the revenue generated by winter production.

HEATING EFFICIENTLY

When it comes to reducing heating costs, the most significant factor is how airtight the greenhouse is. Fresh air entering through a tear or side opening that was closed improperly can quickly cause a spike in heating bills. One of the best strategies for overall energy efficiency is simply to regularly inspect your structure. A low-cost strategy indeed!

The overall volume of air within the shelter will also affect its energy efficiency. A large air mass will cool much more slowly than a small air mass. A bigger greenhouse will retain heat better than a small shelter.

Energy losses are also highest near openings and junctions (doors, where the walls meet the ground, ends, etc.). This is why structures with a greater height and footprint will more effectively retain heat. It also explains the energy efficiency of multi-span greenhouses, in which junctions represent a smaller fraction of the total surface area.

Another good strategy to ensure energy efficiency is to install a modern and efficient furnace, which should include perforated tubes connected to the heater and a squirrel cage (fan), to properly distribute warm air throughout the greenhouse. The squirrel cage pushes air into the

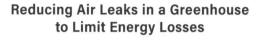

Reducing Air Leaks in a Greenhouse to Limit Energy Losses

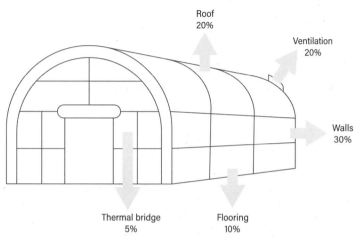

Roof
20%

Ventilation
20%

Walls
30%

Thermal bridge
5%

Flooring
10%

tubes, which run along the beds to distribute heat right at the base of the plants. This method is particularly effective in winter, bringing warm air to crops that grow close to the ground, like lettuce and spinach.

In our experience, all of these strategies are essential for winter production. But in addition to striving for energy efficiency, growers must learn how to synchronize temperatures with the crops' growth cycles. In winter, the greenhouse should be heated according to daylight hours. Vegetable growers must provide warmer temperatures on sunny days, since the combined effect of light and heat creates optimal conditions for photosynthesis. However, on cloudy days, when ideal conditions are unavailable, there's no point in significantly raising greenhouse temperatures. This means, for instance, that heating during February and March, when day length exceeds ten hours, will generate better yields than in December or January, when sunlight hours are at their lowest.

Some greenhouses are unheated throughout winter because they contain vegetables that are very cold hardy. Even these highly resilient crops, however, can reach a breaking point, suffering irreversible damage during a hard freeze. To avoid losses, the safest bet is to establish a plan that includes a preventive heating source. This emergency heating can be turned on before temperatures drop below 22°F (−6°C) under the row cover for more than three consecutive days. To break the cycle of a prolonged freeze, the greenhouse must be heated to above 46°F (8°C) for six to seven hours. The cost of this intervention can be justified by the even greater financial consequences of losing a crop.

Maintaining Continuous Air Circulation

Just as supplemental heat can positively affect overall yield and quality of winter greens, managing relative humidity is as critical. In winter, humidity is particularly high because cold air is more likely to be saturated with water than warm air. The relationship between relative humidity and temperature is inversely proportional: as the air cools, relative humidity increases, and vice versa.

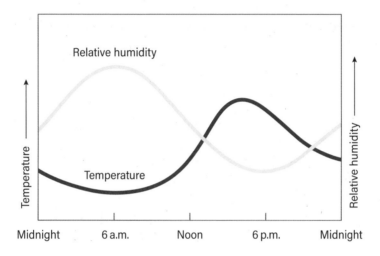

High relative humidity creates problems in winter as it provides optimal conditions for fungal diseases like downy mildew, powdery mildew, and pythium root rot. These can easily spread in a greenhouse and cause major crop decline or even crop failure. A winter grower must intervene, ideally without resorting to spraying a biofungicide. Our main strategy is to dehumidify our greenhouses by following a simple rule of thumb: we don't allow the relative humidity to exceed 90 percent for more than a few consecutive days. We achieve this by heating the greenhouse for less than an hour, while slightly raising the side openings to remove moist air and create a drier climate. To implement this strategy, you must clear snow off the sides of the greenhouse and, most importantly, install side openings a few feet above the ground when building the greenhouse. This last detail will ensure that side openings are not obstructed by snow, which would make using them impossible.

It is also possible to dehumidify an unheated shelter by using the heat from the sun. To do this, simply ventilate the greenhouse when the sun is shining by opening the door or the sides.

The other, more practical option is to install a positive-pressure ventilation system that draws in outside air, bringing a fresh supply into the greenhouse. The air then mixes at the peak before reaching the plants, generating a more uniform climate in the greenhouse.

The Importance of Continuous Airflow

Good airflow in a greenhouse is also crucial for winter growing. Because warm air rises and cool air falls, cooler air tends to accumulate and stagnate around the crops. Adequate air circulation helps avoid this by keeping colder air circulating, thus raising temperatures around the plants. The best way to achieve this is by simply installing horizontal airflow fans (HAF) and running them 24/7 in winter. They are inexpensive to run and well worth the investment. Our venting tubes also help in this way, which is why we run them 24/7 even when no heat runs through them.

Our greenhouses also have openings in the roof, to allow for additional ventilation when needed. On sunny days, even in winter, the inside temperature can rise dramatically. That's why it is essential to have accessible openings in greenhouses, like roll-up sides and roof vents. As temperatures rise, these must be used to release hot air and avoid crop stress, because excess heat can cause many winter crops to bolt, resulting in considerable losses.

Rethinking How We Heat Greenhouses

So far, we've had to use propane to efficiently heat the greenhouses at Ferme des Quatre-Temps. The main reason that propane is so widespread on small farms is simple: the cost of equipment and installation is largely lower than other options. Unfortunately, this energy source is non-renewable, and the monthly bill is expensive. But given the alternatives currently available on the market, we don't have much of a choice.

Why not run the greenhouses on electricity? Or tap into geothermal energy? Or biomethane?

In assessing alternatives to fossil fuels such as propane, the key considerations have to be profitability over ecology, or any other ideological aspects. The margins in our line of work are too small to even consider otherwise. A winter farming project that didn't make any profit, or worse, operated at a loss, wouldn't last very long or serve anyone.

However, for most of us small-scale growers, environmental awareness is not only a concern, it's a guiding value, especially in our climate change era.

While researching for the best alternative to carbon-neutral sources, we've selected the solutions with the most potential:

- Electrical heating (electric coil and heat pump)
- Geothermal heating: the climate battery example
- Radiant heat

Each alternative has pros and cons, but all options are worth exploring as potentially permanent solutions to eliminate fossil fuels in heating greenhouses. You will see that no solution is sufficient on its own: you must always combine two or more to build a reliable and efficient heating system.

Going Electric: Could Be the Best Alternative

The option of heating greenhouses with electricity has long been avoided because of the high cost of heaters and power. In many parts of the world, electricity is not locally sourced and is sometimes produced by coal turbines or nuclear plants, two options that aren't environmentally friendly.

In Quebec, our situation is different in that electricity is produced locally and, most importantly, generated by hydro dams, a relatively carbon-neutral solution. This electricity is publicly owned, with an imperative to keep prices as low as possible for the common good. This is the favorable context in which we conduct our research at Ferme des Quatre-Temps.

Heating a greenhouse with electricity may seem like a complex and technical process at first, but it really isn't. To understand this system, you must know that this choice involves either installing a furnace with an electric coil or using a heat pump. These two options are radically different, and comprehending that difference is key.

Before you embark on an electric venture, whether a furnace or a heat pump, first verify that you have access to an electrical input called three-phase.

Power required to heat a greenhouse with electricity:
600 volts — three-phase
200–400 amps

Beware: If the three-phase power entrance isn't already set up at your farm, the installation can be quite expensive and can use a large portion of your budget.

Quick tip: To tell if your farm has access to three-phase power, look at the power lines on your street; there should be three electric lines per pole.

Resistive Heating

This electric system is the simplest one available: its operation is similar to an electric calorifier used to heat our houses. Current is sent through a conductive material that creates electrical resistance. The result is dissipation of energy in the form of heat. The magic of this system is that there is no loss of energy: 1 kWh electrical = 1 kWh thermal.

In a greenhouse, this heating method is used in the form of an electric coil combined with a propane heater as a backup. Both furnaces (electric and propane) work on the same pipes and fans.

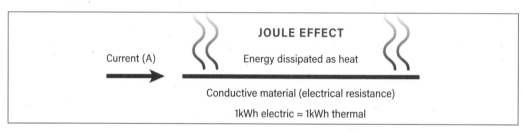

JOULE EFFECT

Current (A)

Energy dissipated as heat

Conductive material (electrical resistance)

1kWh electric ≈ 1kWh thermal

Dual-energy heating system

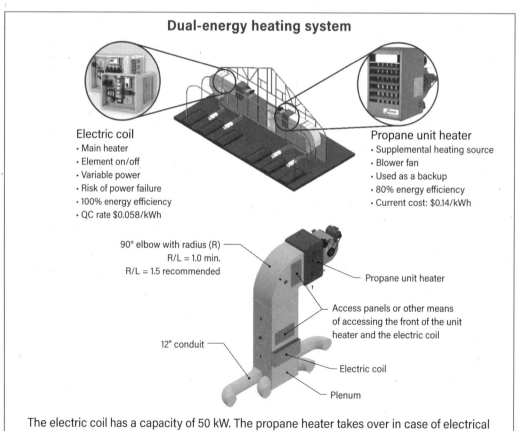

Electric coil
- Main heater
- Element on/off
- Variable power
- Risk of power failure
- 100% energy efficiency
- QC rate $0.058/kWh

Propane unit heater
- Supplemental heating source
- Blower fan
- Used as a backup
- 80% energy efficiency
- Current cost: $0.14/kWh

90° elbow with radius (R)
R/L = 1.0 min.
R/L = 1.5 recommended

Propane unit heater

Access panels or other means of accessing the front of the unit heater and the electric coil

12" conduit

Electric coil

Plenum

The electric coil has a capacity of 50 kW. The propane heater takes over in case of electrical failure. For this system to work, two propane units are needed with a capacity of 58 kW each.

Heat Pump? It Might Be the Best Option Out There

A heat pump similar to one used in a house can be installed to heat a greenhouse. The main advantage is that it is more energy efficient than all alternatives; it uses very little energy to create a lot of heat. It's also a very serviceable, widely available solution.

How a heat pump works is simple: heat moves from hot to cold, and through its coil, the pump reverses this process to transform cold air into hot air. In our case, installing a heat pump in a greenhouse enables us to take the outside cold air, transform it into hot air, and dissipate it in the greenhouse.

Several heat pump models exist, but the one that is best suited to small farms is the air-to-air system, which captures cold air and transforms it into hot air. Its main advantage is its low cost, making it relatively accessible, although it is important to mention that heat pumps are more expensive than resistive electric systems.

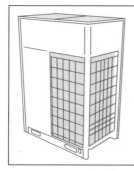

A significant downside to using a heat pump for a greenhouse is that its performance decreases with very cold temperatures, when it is needed the most. For that reason, you absolutely need a backup heater.

Surface Geothermal Heating: The Climate Battery Example

Another highly appealing alternative is near-surface geothermal energy, which allows heat from the ground to be recovered and distributed throughout a greenhouse. As this technology is becoming more widely available, installation costs are decreasing. This is good news and, honestly, something to be excited about.

The principle of a climate battery is to capture solar energy and store excess heat in the soil. It draws warm air into agricultural drains shallowly buried under large rocks near the greenhouse and then releases it inside the greenhouse. At the time of doing our research, installation costs were about $10,000 per greenhouse, about one-tenth that of a regular geothermal system. A climate battery also requires a dual-energy heating system, which means investing in another heating solution. However, what's really interesting about this system is that heating a greenhouse is practically free. You pay only for the energy needed to activate the fans that draw warm air inside the greenhouse.

Surface geothermal heating is a very attractive option, but its exact power (how many Kwh it produces) is still unknown. At Ferme des Quatre-Temps, we are currently trialing to determine the cost and size of the system and the exact setup where it should prove to be effective and profitable. We know that most systems wouldn't be powerful enough to heat a greenhouse from –4°F to 68°F (–20°C to +20°C), but since this is not our objective, we are hopeful. Keeping a greenhouse all winter to around 37°F to 41°F (3°C to 5°C) is our goal, and if surface geothermal heating could do it at a low installation cost, this could revolutionize winter growing. We could minimally heat our winter crops at almost no production cost.

Surface geothermal heating system

Two very large ducts, 24-inch (60 cm) diameter, are buried outside the greenhouse about two feet under large rocks. First, make sure the water level is below the ducts.

Two very powerful fans are installed inside the greenhouse at the base of the ducts. They suck the air from outside, pulling it inside the ducts where it's heated and released into the greenhouse.

Greenhouse

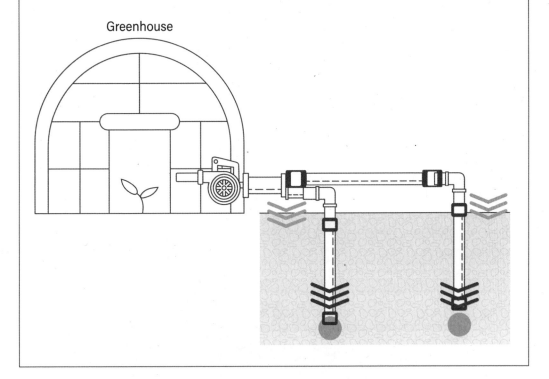

Underground Radiant Heating

Unlike conventional greenhouse producers who usually grow tomatoes and cucumbers in containers using hydroponic systems, most organic greenhouse growers plant their crops in the ground. This is what organics stands for and what the organic label should entail, in our opinion, even though in some countries, such as the USA, hydroponics producers can receive the organic certification.

Thus, any heating solution designed to meet organic standards can take this into consideration. Soil is part of the equation. Why not turn this into an advantage? Soil is an immense thermal mass that can be heated, even in the absence of sun. Once warmed, this thermal mass tends to retain heat, which can stabilize the ambient temperature of a large greenhouse. This is because, unlike air, soil has the capacity to store the energy it receives. Heating the ground rather than the air is a promising avenue. Moreover, plants can withstand colder air temperatures when their roots are warm.

This idea is far from new. In fact, many greenhouses where tomatoes grow in the ground year-round are already equipped with this kind of underground heating system. Installation involves burying tubing 18 inches (45 cm) deep and connecting it to an electric boiler that heats the liquid (food-grade glycol) circulating in the tubes, which are buried under the crop. At Ferme des Quatre-Temps, we have a similar system in a large multi-span greenhouse. We usually turn it on in March to make sure that the roots of our tomato seedlings will stay warm when we transplant them.

We are just beginning to experiment with this system in winter. As these are our first attempts, it's too soon to definitively recommend it for winter growing, but so far, we have seen a positive effect on crop yields.

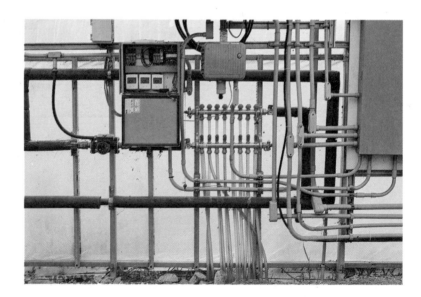

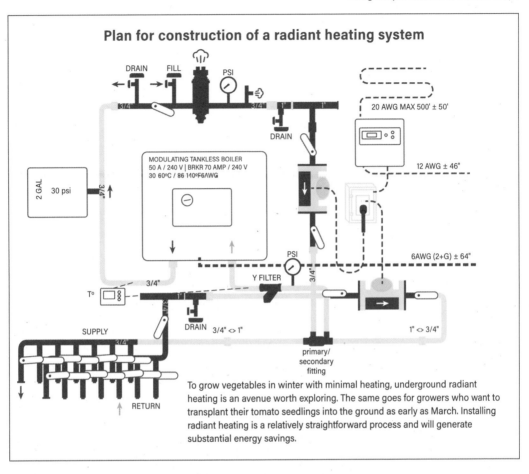

Plan for construction of a radiant heating system

DRAIN FILL PSI

3/4" 3/4" 1" 1"

20 AWG MAX 500' ± 50'

DRAIN

MODULATING TANKLESS BOILER
50 A / 240 V | BRKR 70 AMP / 240 V
30 60ºC / 86 140ºF6AWG

12 AWG ± 46"

2 GAL 30 psi

3/4"

6AWG (2+G) ± 64"

PSI

3/4"

Tº

3/4"

Y FILTER

SUPPLY

3/4"

DRAIN 3/4" <> 1"

1" <> 3/4"

RETURN

primary/
secondary
fitting

To grow vegetables in winter with minimal heating, underground radiant heating is an avenue worth exploring. The same goes for growers who want to transplant their tomato seedlings into the ground as early as March. Installing radiant heating is a relatively straightforward process and will generate substantial energy savings.

The best way to distribute heat throughout a greenhouse is to use perforated tubes that send warm air underneath the row covers. This keeps the heat close to the plants, rather than losing it to the ambient air.

53

Comparing Greenhouse Heating Strategies

Type of heating	Pros	Cons
Climate battery (one form of geothermal energy) Initial cost*: $10,000	· Zero energy cost other than the electricity needed for the fans · Carbon-neutral · Simple and cheap to install	· Not powerful enough to heat a greenhouse yearlong at 42°F (5°C) · Technology not proven yet, still in development phase · Not a reliable heating system
Resistive electric heating Initial cost: $40,000 (including backup propane heaters)	· Low operating costs in Quebec, due to preferential rates offered for greenhouse production · Carbon-neutral · Energy efficiency near 100% · Low maintenance	· Requires three-phase power on the farm · Expensive to install · Not environmentally friendly if electricity is generated with a nonrenewable resource (coal, nuclear)
Electricity: Heat pump Initial cost: $55,000 (including backup propane heaters)	· Lower operating costs in Quebec, due to preferential rates offered for greenhouse production · Carbon-neutral · Energy efficiency over 100% · Low maintenance · Has the potential to cool a greenhouse in hot weather and dehumidify it at all times	· Requires three-phase power on the farm · Very expensive to install · Not environmentally friendly if electricity is generated with a nonrenewable resource (coal, nuclear)
Geothermal Initial cost: $125,000	· Environmentally friendly · Low cost of heating once the system is installed · Carbon-neutral	· Highly expensive to install · Not yet fully adapted to the reality of greenhouses; lack of expertise among installers · Requires another heating system with a dual-energy mode
Propane (and natural gas) Initial cost: $25,000	· Simple and inexpensive to install · Easy to procure · Some furnaces are now at 93–97% efficiency	· Combustion emits pollutants · Propane is costly · Prices will fluctuate · Nonrenewable resource
Wood Initial cost: $15,000	· Inexpensive · Carbon-neutral	· High labor requirement (e.g., putting wood in the furnace at night)
Fuel oil and heating oil Initial cost: $15,000	· Low maintenance	· Expensive · Very unclean emissions · Low energy efficiency · Nonrenewable resource

* The costs listed in this table are estimates whose purpose is to give an order of magnitude and not an exact cost.

Winter Crop Planning ❄

Organic farming appealed to me because it involved searching for and discovering pathways, as opposed to the formulaic approach of chemical farming. The appeal of organic farming is boundless; this mountain has no top, this river has no end.

— Eliot Coleman

As shown in the previous chapter, the use of simple shelters not only protects crops from bad weather but also creates conditions that promote healthy plant growth. While the cold is certainly one factor limiting winter growing, it is not the only one nor is it the hardest to work around. For vegetable growers operating in a northern climate, one of the biggest hurdles to overcome is, in fact, the decrease in daylight hours and solar energy experienced in November, December, and January. For a successful winter vegetable production, anyone embarking on this adventure needs to understand the ins and outs of the season, and work differently than in summer.

To develop that exceptional flavor and grow healthy winter crops, two essential steps are required: developing strong seedlings, then hardening off these young plants so they can withstand winter conditions.

In his *Winter Harvest Handbook*, Eliot Coleman, the pioneer of winter growing, describes how important it is to observe nature to better understand it. And he's right. As soon as the first light frosts appear in the fall, cold-hardy plants start concentrating sugars in their cells. This is the beginning of their acclimation process, a preparatory stage that will allow them to survive colder nights at the start of winter. Sugar water in plant cells acts like antifreeze. This is one of the most extraordinary survival mechanisms seen in the plant world. To visualize the phenomenon, just try putting fruit juice next to a bottle of water in the freezer. The water will solidify much sooner than the fruit juice. In sweet cells, the freezing point is lower, which increases a plant's frost resistance. This is because when liquid freezes inside the cells, it forms ice crystals that eventually tear through cell membranes, causing the plant to die.

The way in which plants react to the cold is something we can use to our advantage. Winter vegetables not only have a better chance of surviving a hard freeze, they also develop quite a sweet taste, due to the sugar concentration in their cells. The result is a fantastic flavor that charms anyone who has had the pleasure of sampling winter produce. This is a powerful selling point.

Freeze Damage in Cells

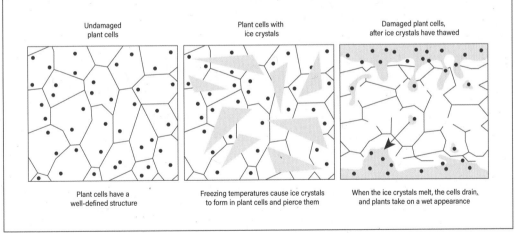

Undamaged plant cells	Plant cells with ice crystals	Damaged plant cells, after ice crystals have thawed
Plant cells have a well-defined structure	Freezing temperatures cause ice crystals to form in plant cells and pierce them	When the ice crystals melt, the cells drain, and plants take on a wet appearance

Planting Dates: Planning for Diminishing Day Length

The biggest limiting factor for winter crops is a lack of light, and not a lack of heat. In the fall, plant growth gradually slows down in almost all crops. As soon as day length drops below ten hours, growth essentially grinds to a halt. And so, every year, we witness a race against the clock on all farms that have a winter vegetable production. To make sure that vegetables will be close to maturity before they shut down for winter, each farm has to determine the best time to pull out summer crops and plant winter crops. In this game, time is not on our side. We must find a balance between getting those last harvests out of summer plants while still preparing winter crops. For abundant winter harvests, plants need to have reached at least 70 percent of their growth before day length drops below ten hours. At Ferme des Quatre-Temps, in Hemmingford, this critical moment arrives around November 1. After that, plants can still be harvested, though they will grow extremely slowly, if at all. The greenhouse or high tunnel then becomes a big cold room from which we draw fresh vegetables.

When Day Length Is Less than 10 Hours
Examples for Different Cities

City	Dates when day length is less than 10 hours
Portland, USA	November 3 — February 6
Burlington, USA	November 5 — February 5
Washington, USA	November 16 — January 24
San Francisco, USA	November 20 — January 20
Dallas, USA	December 16 — December 26
Montréal, Canada	November 1 — February 1
Quebec, Canada	October 31 — February 1
Paris, France	October 29 — February 13
Toulouse, France	November 6 — February 5

Starting in February, the days lengthen significantly. From then on, plants emerge from a winter sleep and start to grow vigorously again, in shelters protecting them from the drying effect of cold winds.

Several resources are available to determine the dates when day length is less than ten hours, according to your location. With this date in mind, you can use the days to maturity (DTM) for each crop to count backwards and decide when they should be planted in a shelter. It's essential to add a few days to this date, to account for the diminishing sunlight and cold temperatures, which both slow down growth as autumn turns into winter. As a result of these conditions, the DTM for winter vegetables are not the same as their summer DTMs. From season to season, note-taking on planting and harvest dates will allow you to adjust DTMs for each of your winter crops.

Here are examples for a few crops, to help you understand how to calculate transplant or direct seeding dates according to DTMs. In these calculations, the cut-off date that lines up with the last 10-hour day is November 1st.

Crop	Experimental days to maturity (DTM)	Number of days to reach 70% maturity	Target date for direct seeding or transplant
Tatsoi	35	25	October 7
Spinach	45	32	September 30
Turnip (Hakurei)	60	42	September 30

Note that the number of days to maturity can increase and even double depending on temperatures and sunlight. Adjusting your DTMs for each crop, according to your farm's conditions, is essential.
The target date selected using this calculation is theoretical and will vary from one crop to another. This is still a good starting point, which you can then change according to your experiments.

Adjusting Planting Dates for a Continuous Harvest

For many vegetables that are harvested only once, determining the right planting date is relatively simple. You need to make sure that the plant will have almost reached maturity before day length drops to ten hours or less. However, this is not the case for vegetables that will be harvested multiple times over the winter (known as cut-and-come-again crops), like Salanova lettuce, arugula, and Asian greens. In a winter context, where growth comes to a near standstill, we can only get three cuts out of such crops over the entire season.

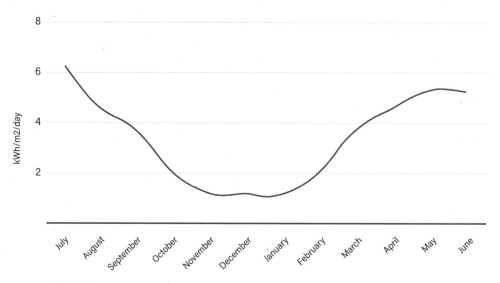

Received light (kWh/m²/day) over 12 months

By finding the light received according to the time of the year, it is possible to graph this data as above. The data represent the light received at Ferme des Quatre-Temps in Hemmingford, Quebec (45° north).

This graph is very useful for planning a winter production, and we recommend that you produce a similar one for your location.

The idea is to synchronize your production with the light received. In our case, two periods of plant growth are possible in winter: #1, September 15 to November 1; #2, January 15 to April 1.

To secure a constant supply of these vegetables, you must adapt your seeding strategy according to the space available in your shelters. The goal is to have two or three different planting dates for these crops, allowing seven to ten days between seeding each succession. For us, this method increases the odds of successfully growing a continuous harvest and improves the predictability of our vegetable offering.

To choose planting dates for greens like tatsoi, arugula, mustard, and all the other components of our winter mesclun, we consider the fact that their leaves must be harvested when they are no bigger than the palm of a hand. This size is what makes a mesclun, and it is the most important factor in determining their maturity.

In summer, greens take ±30 days to reach the size we need. Based on our experience, tatsoi seeded on October 5 will mature in roughly 35 days, which is about the time required in summer. However, for a second succession seeded on October 15, tatsoi DTMs jump to at least 50 days. When the planting dates are delayed in the fall, we've observed a clear trend, with a drastic increase in DTMs as the pivotal November 1 date approaches.

Days to Maturity Increase as Sunlight Diminishes

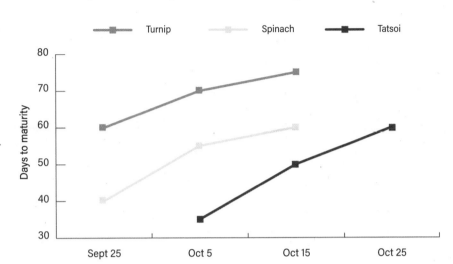

This means that the longer we wait to seed a crop in October, the more its DTMs will be. When making a winter crop plan, you must take this reality into account. The graph above illustrates this increase in DTMs seen in the fall. After a first winter of experiments, you should be able to draw a similar graph for all your crops. This tool will be essential for planning your next winter season.

By now, you've probably realized that there really is an art to determining the right planting dates so that crops will overlap smoothly, providing an appealing and continuous vegetable offering throughout winter. At Ferme des Quatre-Temps, we are fortunate to have access to several shelters, so space is not a limiting factor. We make sure to plant our winter crops in shelters as soon as a spot is freed up by a summer crop. The planting date is therefore well before that pivotal moment when day length dips below ten hours. This ensures a smooth transition from one crop to the next, mitigating the effects of diminishing sunlight over the days leading up to the winter solstice.

Adjusting planting dates is a process of continuous improvement, year after year. Unfortunately, there is no magic formula that can be applied to all farms. Experimentation and rigorous note-taking are the keys to determining what works best in your context. You can also follow the schedule of another well-established northern farm and make changes along the way. This is how we started our experiments at Ferme des Quatre-Temps. As an example, here is a table of our own dates for fall and winter plantings. Of course, these dates are for a farm in southern Quebec, but they can still serve as a guide.

Winter Planting Dates

Crop	Outside, under P19 cover	Caterpillar tunnel	High tunnel	Greenhouse, minimally heated
Arugula	September 1	September 15	October 1	October 15
Asian greens	September 1	September 15	October 1	October 15
Baby kale	September 1	September 15	October 1	October 15
Baby Swiss chard	—	Mid-August	September 1	September 15
Bok choy	September 1	September 25	October 1	October 1
Carrot	August 1 No need for P19	August 10	March 15	February 15
Celery	—	July 15	August 1	August 30
Chinese cabbage (Tokyo Bekana)	September 1	September 25	October 5	October 15
Cilantro	September 15	September 25	October 5	October 10
Claytonia	August 20 No need for P19	September 10	September 15	September 20
Dandelion	August 1	August 15	September 1	September 15
Green onion	August 15	August 15	August 15 or February 15	August 15 or February 15
Kale (mature)	August 15	September 1	September 15	October 1
Komatsuna	August 1	August 15	September 1	September 15
Lettuce (Salanova)	August 15	September 1	September 15	October 10
Mâche	August 15	September 1	September 15	September 15
Mustard	September 1	September 15	October 1	October 15
New potato	May 1	April 15	April 1	March 15
Parsley	July 15	August 1	August 1	August 30
Radish	September 1	September 15	September 15	September 20 or mid-February
Senposai	August 1	August 15	September 1	September 15

Crop	Outside, under P19 cover	Caterpillar tunnel	High tunnel	Greenhouse, minimally heated
Sorrel	July 15	August 1	August 1	September 1
Spinach	September 1	September 15	September 15	October 1
Swiss chard (Mature)	August 15	September 1	September 15	October 1
Turnip	September 1	September 15	September 15	September 20 or mid-February
Watercress	August 15	September 1	September 15	October 1

Northern countries don't all experience the same northern reality. Their geographical location along the Earth's axis will dictate, to a significant extent, prevailing conditions. Countries located along a parallel near the equator will thus be exposed to more sunlight. For cities in the far north, it is quite the opposite. Whitehorse, Yukon (Canada), and Oslo, Norway, are located near the 60th parallel, where in winter there are 18-hour nights and 6-hour days. With so little sunlight, it's impossible to grow anything in this period without artificial lighting.

In comparison, Montréal is located along the same parallel as the city of Bordeaux, France. For both cities, average day length is nine hours and twenty-seven minutes over December, January, and February. This means that, given equal temperatures, what grows in winter in Bordeaux should grow just as well in southern Quebec.

This is the rationale that spurred Eliot Coleman to visit vegetable growers in Bordeaux, to see what was being harvested in winter. He went on to use the knowledge and techniques he had observed in France to run his own experiments in the greenhouses and high tunnels on his market garden, located in Maine, in the northeastern United States.

Decreasing Crop Density to Promote Better Growth

Once our planting dates are chosen, there are a host of other factors to consider to create optimal conditions that will allow plants to reach maturity. Crop spacing is one variable that can be adjusted to compensate for decreasing sunlight. By increasing both the distance between rows and the number of plants per row, we ensure better light penetration and airflow. According to the biointensive model that we follow on our farm, crops are usually planted much more densely than on mechanized farms. One of the benefits of density is that crop foliage grows close together, obstructing the light supply that would otherwise go to weeds. This allows us to generate better yields per square meter than the standard yields on mechanized farms.

But in winter, as sunlight decreases, it's important that all the leaves of each plant still have access to optimal sunlight so they can photosynthesize, which is essential for strong development. This requires some adjustment; for instance, most of our crops that are seeded at ten rows per bed in summer drop down to six rows in winter.

For salad greens, radishes, and turnips, we decrease the number of rows in the bed and increase their density within the row. In our experience, the best practice is to sow six rows at about 70 percent of the seeding rate assigned to ten-row summer seedings. For example, in summer we sow a total of 3.4 ounces (95 g) of arugula seed in a ten-row bed. In winter, we reduce the quantity and sow about 2.5 ounces (70 g, or 3.4 oz. × 70%) on a six-row bed. For more information, see Appendix 1: Winter Crop Spacing (p. 236).

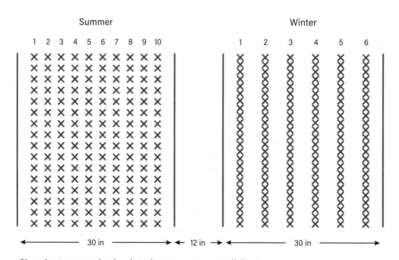

Changing crop spacing in winter increases access to light.

Preparing Plants for Winter

For a successful winter production, another important factor is the need to acclimatize plants to the cold. As explained earlier, plants concentrate sugar in their cells in winter to protect themselves from the effects of low temperatures. To effectively set in motion this defense mechanism, it's crucial that they be prepared (hardened off) well before and after planting, and to provide gradual exposure to the cold. When these steps are taken, many crops will become very frost resistant. We have stopped counting the number of times we've stepped into an unheated greenhouse on a winter morning and opened a row cover to reveal a bed of fully frozen vegetables after a very cold night. For a vegetable grower who has little experience in winter farming, this can be shocking—it's a complete loss! But as the morning goes on, the sun's energy slowly warms the greenhouse, and a miracle occurs: the plants thaw and become as beautiful and alive as the day before. Witnessing this metamorphosis is extraordinary, but it is also a daily occurrence on a farm with crops that have been hardened off.

Several crops or varieties naturally cope well with the cold. Some can even be quite resilient when faced with a short-lived hard freeze. But these same crops will suffer permanent damage and even die if below-freezing temperatures persist. This is the case for arugula leaves that can endure temperatures as low as 18°F (–8°C) for less than two hours but will suffer permanent damage if exposed to 25°F (–4°C) for more than twenty-four hours. Arugula, like all cold-hardy crops, has an energy reserve that will bring it back to life as soon as the sun warms the greenhouse. That said, plants will more effectively recover from temperature variations that are typical of winter growing if they have been hardened off well before freezing temperatures strike. To achieve adequate cold tolerance, we recommend following two guiding principles.

PRINCIPLE NO. 1: DEVELOP A STRONG ROOT SYSTEM

In order to withstand temperature changes, winter crops must have well-established root systems. To make sure our plants develop strong root systems well before temperatures drop below freezing, we need to act fast. Once summer crops are removed from shelters, winter crops are quickly transplanted or seeded into that same space, to take advantage of the favorable fall weather. Most of these have been growing in our nursery to get a head start of a few weeks. The longer plants can benefit from mild temperatures and long daylight hours, the stronger their root systems will be. Important detail: when preparing our winter seedlings in the nursery, any crop that will eventually be transplanted is seeded into larger containers. So, for example, a vegetable that would be sown into a 128-cell tray in summer is instead seeded into a 72-cell tray for winter. This gives seedlings more space to develop a robust root system and allows us to give winter crops a head start while keeping summer crops in the greenhouse for one or two more weeks.

The Nursery

The nursery is a greenhouse adjacent to our conditioning room (insulated north wall) where we start all the seedlings that are later transplanted in the field or in the greenhouse.

The nursery is equipped with mobile tables, a heating system, and an opening roof for ventilation. We use only natural light for the growth of the plants.

Seven days before seedlings are transplanted, we move them into the hardening-off area that is more open to the outside, which slowly prepares them for more difficult conditions.

PRINCIPLE NO. 2: HARDEN OFF PLANTS GRADUALLY

To help plants withstand more severe frosts, it's critical to harden them off in an intentional and gradual manner. We purposely expose crops to colder temperatures by leaving them without a row cover or by opening the greenhouse or tunnel to let the cold air in. This occurs around mid-November, when most of the planted crops are established and have almost reached maturity. We gradually increase their cold tolerance before temperatures begin to drop below freezing. Plants that have not been hardened off, in comparison, are more affected by below-freezing temperatures. Ice crystals form in their cells and tear through them, resulting in a dark green color and a wet appearance. The damage is irreversible, and the crop will unfortunately be lost.

Example of a plant in which cells have not been damaged by ice crystals

Quick Tip

Brassicas (tatsoi, bok choy, arugula, etc.) tend to bolt in February and March. To prevent this, try to keep temperatures below 68°F (20°C) and plan to harvest these crops before the beginning of March.

The Importance of Note-taking…And Following an Example

Growing crops in winter requires patience and curiosity. You need to determine seeding dates, adjust the number of days to maturity, decide how many successions to plant so you can harvest at the right time, understand how to space plants appropriately, and estimate the right timing to harden off plants…. All this makes managing a winter season quite complex. To ensure success, observation and note-taking are crucial. Rigorous documentation is the key to understanding this web of factors affecting production.

By monitoring temperature, growing practices, sunlight, and other factors you will better understand how your crops, soil, and shelters react to the cold. Establishing serious note-taking procedures is a must.

The following data should always be included in your notes:

- Seeding and transplant dates for each crop
- Harvest dates for each crop (to determine their respective days to maturity)
- Number of cuts made in each bed of greens, and the number of days between cuts.
- Yields per crop compared to heating costs (if applicable)
- Temperature at which a crop is killed by the cold, and the context (number of row covers, heated or unheated space, etc.)
- Date when diseases or pests appear, and actions taken

Example of Note-Taking for the Winter Season

Crop	Location	Seeding or transplant date	Harvest Date	Yield lb./bed (kg/bed)	If crop was killed by cold, record temperature °F (°C)	Diseases and insects, first sighting (date and action)
Arugula	Greenhouse No 1	October 1	November 5	55 lb. (25 kg)	14°F (−10°C)	—
			December 15	37 lb. (17 kg)		
Lettuce	Greenhouse No 2	Sept 15	November 1	33 lb. (15 kg)	—	Powdery mildew – December 1
			November 15	22 lb. (10 kg)		
			December 1	20 lb. (9 kg)		
Spinach	…	…	…	…	…	…

Striving to Improve with Each Season

Growing vegetables in winter is certainly not an exact science, and every context is different. Even with examples like those shared in this book, there will always be a need for experimentation to understand your own farm. Data collected over the seasons is essential for creating an annual crop plan that is suited to your reality.

Since winter production is gaining in popularity, it seems to us that we need to leverage our collective skills and knowledge to shorten the individual learning curve on each farm. We believe this is critical. Agronomic research chairs and projects like Quebec's Centre d'expertise et de transfert en agriculture biologique et de proximité (CETAB+) are immensely important and contribute to improving farming practices.

Photo courtesy of CETAB+.

The Centre d'expertise et de transfert en agriculture biologique et de proximité (CETAB+) has established a technological showcase to study cool weather crops grown under cover. They test and demonstrate winter vegetable production then collect and share results with the farming and agronomy community. They have also studied several farms operating in winter, which has strengthened and solidified their results.

When the results of such experiments are shared, producers are able to avoid costly mistakes and address their winter production well ahead of schedule.

The planting calendar presented below is the result of experiments run over several seasons. It has been designed to yield a diversified offering for both restaurants and winter baskets, sold to customers who delight in our range of winter vegetables.

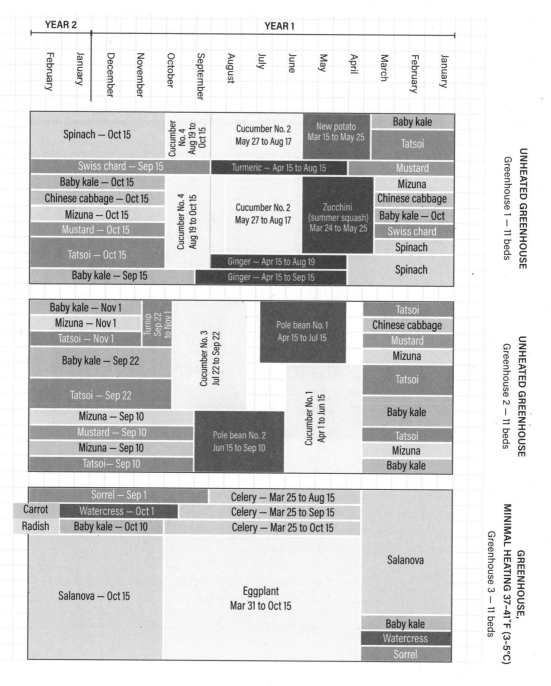

GREENHOUSE, MINIMAL HEATING 37–41°F (3–5°C)

Greenhouse 3 bays — 24 rows

	YEAR 1											
January	Swiss chard	Swiss chard	Kale and bok choy		Celery	Arugula	Chicory	Spinach	Salanova	Arugula	Chinese cabbage	Parsley
February												
March												
April	Tomato — Mar 15 to Oct 1	Tomato — Mar 15 to Oct 1	Tomato — Mar 15 to Oct 1	Tomato — Mar 15 to Oct 1	Tomato — Mar 15 to Oct 1	Tomato — Mar 15 to Oct 1	Tomato — Mar 15 to Oct 1	Ginger — Apr 15 to Aug 20		Tomato — Mar 15 to Oct 25	Tomato — Mar 15 to Oct 25	Tomato — Mar 15 to Oct 25
May												
June												
July												
August												
September												
October	Kale and bok choy Oct 1	Swiss chard and green onion Oct 1	Swiss chard and green onion Oct 1		Spinach — Oct 10	Chicory — Oct 10	Salanova — Oct 10	Celery — Aug 20	Parsley — Aug 20			
November										Arugula	Arugula	
December												
YEAR 2 January									Chinese cabbage			
February												

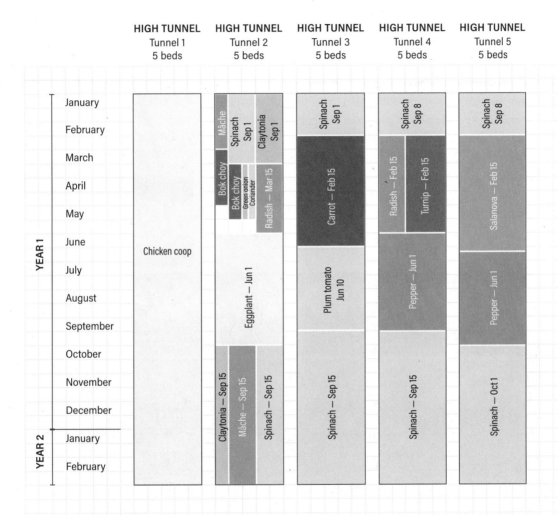

HIGH TUNNEL
Tunnel 1
5 beds

HIGH TUNNEL
Tunnel 2
5 beds

HIGH TUNNEL
Tunnel 3
5 beds

HIGH TUNNEL
Tunnel 4
5 beds

HIGH TUNNEL
Tunnel 5
5 beds

YEAR 1

January
February
March
April
May
June
July
August
September
October
November
December

YEAR 2

January
February

Chicken coop

Mâche

Spinach Sep 1

Claytonia Sep 1

Bok choy

Bok choy

Green onion

Coriander

Radish — Mar 15

Eggplant — Jun 1

Claytonia — Sep 15

Mâche — Sep 15

Spinach — Sep 15

Spinach Sep 1

Carrot — Feb 15

Plum tomato Jun 10

Spinach — Sep 15

Spinach Sep 8

Radish — Feb 15

Turnip — Feb 15

Pepper — Jun 1

Spinach — Sep 15

Spinach Sep 8

Salanova — Feb 15

Pepper — Jun 1

Spinach — Oct 1

Crops sown later in the season (after September 30) have this advantage: long after Christmas, once the sunlight begins to increase in February, they can provide excellent harvests.

If you have enough space in your shelters, the best strategy is to sow winter crops every week from the beginning or middle of September until the end of October. This approach will result in good harvests before Christmas and in the spring, until summer crops are planted in the greenhouse.

The dates suggested in the table on pages 62 and 63 represent our tests, run in Hemmingford. What works elsewhere depends on specific climate and—most importantly—latitude, which determines the amount of sunlight available for growing vegetables. The proposed planting dates represent what is realistic, while still considering the end of summer crops (tomato, eggplant, cucumber). It may be possible to plant winter crops earlier on your own farm.

◄ Plants can survive below-freezing temperatures, providing the cold doesn't last too long. Their stored energy will help them recover once they gradually thaw as the sun warms the greenhouse. Crops that are suitable for winter will have a good chance of surviving if their root system is well-established and cold temperatures alternate with warm temperatures. The use of minimal heating also reduces the occurrence of drastic temperature fluctuations.

▲ To harden off plants, the trick is to ventilate the space as much as possible before temperatures drop below freezing. In November, when plants are well-established and temperatures stay above freezing, we keep our greenhouses and tunnels open to gradually expose our crops to colder temperatures.

Vegetables in Winter

❄

Every crop grown in winter has a cold-tolerance threshold; beyond this point, freeze damage will be permanent. For each of these vegetables, here is a summary of the techniques and tricks we use to promote healthy plant growth.

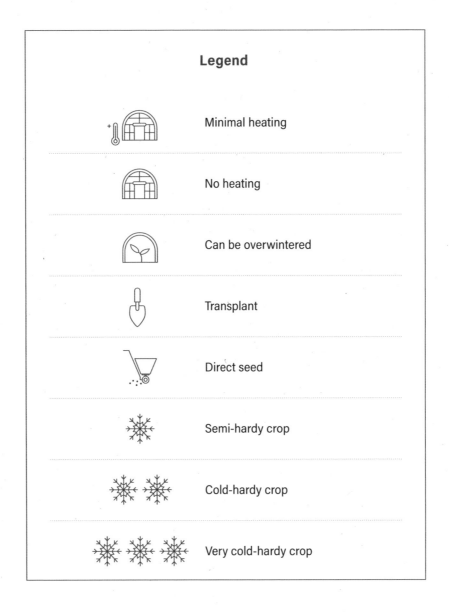

Legend

	Minimal heating
	No heating
	Can be overwintered
	Transplant
	Direct seed
	Semi-hardy crop
	Cold-hardy crop
	Very cold-hardy crop

Semi-Hardy Crops

Temperatures that cause damage: 32°F to 25°F (0°C to –4°C)
Some vegetables, like potatoes and arugula, are quite frost-prone and must be protected from temperatures below 32°F (0°C), to avoid crop losses. They should ideally be grown in a minimally heated greenhouse or with at least one backup heater that can be turned on when the risk of frost is too high.

Cold-Hardy Crops

Temperatures that cause damage: 23°F to 14°F (–5°C to –10°C)
Many cold-hardy crops will benefit from being grown in a minimally heated greenhouse. This heat gain promotes better plant growth and thus a substantial increase in yields. By keeping ambient temperatures above freezing, minimal heating eliminates the need for row covers. And since handling covers requires a lot of time and effort, this represents a significant advantage. The cold-hardy category includes vegetables that we prefer to grow under these conditions, although some can also be grown without heating. Where this is the case, we've indicated the option.

Very Cold-Hardy Crops

Temperatures that cause damage: 14°F to –22°F (–10°C to –30°C)
These cold-hardy vegetables are the pillars of winter production. When protected by a simple shelter like a tunnel or a greenhouse, they can yield diversified harvests throughout the winter, without any heating costs. Exposing these crops to cold temperatures, however, is a limiting factor that significantly slows plant growth.

Arugula

Cultivars

Astro
Esmee

Spacing

6 rows
0.5 inches (1.3 cm)

Seeding date

September 1 to
October 15

Arugula is a leafy green that our customers love, especially in winter when the flavor is milder. The main advantage of arugula is its quick growth. In the fall, the first harvest can happen as soon as 30 to 40 days after seeding. Plant growth slows down during winter, but even a late arugula seeding will quickly start to grow again after January 1, allowing for additional harvests before spring.

For a continuous winter harvest, we plan two or three fall seedings, each 7 to 10 days apart. We sow arugula in six rows, to optimize the amount of sunlight reaching the base of the plants. We harvest arugula using a sharp knife as this maximizes the quality of regrowth. The first cut should be clean and above the growing tip, to allow for a second and third harvest. Dead leaves must be removed with a tine weeding rake after the first and second cuts, for better regrowth and to reduce the risk of developing fungal diseases. For the last cut, a greens harvester can be used to speed up the process.

In colder weather, arugula is one of the first crops to suffer frost damage. To ensure success, it's best to sow it in a minimally heated greenhouse, though it can grow well in an unheated greenhouse. If you choose the latter strategy, protect the arugula on cold nights (23°F [–5°C]) with several layers of row cover, the thickest available on the market. These very heavy covers (1.2 oz./yd.^{2} [40 g/m^2]) will, however, need to be removed on sunny days and pulled back over the crop in the late afternoon, as they allow little light penetration.

Celery

Cultivars

Kelvin
Tango

Spacing

3 rows
12 inches (30 cm)

Start in indoor nursery

70 days in a
32-cell tray

Transplant date

August

The idea of growing celery, like parsley, in winter came from the desire to diversify our winter production. While researching what vegetables to produce in the cold, we recognized the potential of celery. It is easy to grow in winter, but its success depends on one crucial element—its transplant date. We can't stress this enough: it is imperative to transplant celery in August. It is not possible to delay this date.

In order to harvest celery throughout the winter, you must enter winter with mature plants. With the right transplanting date and climate, you are almost guaranteed a successful crop since it is little affected by disease or insects in winter.

We grow celery in minimally heated greenhouses to ensure its survival regardless of the outside temperature and to accelerate the regrowth of stalks after harvest. The main abiotic disorder affecting the crop is black heart, which is caused by a calcium deficiency. We use a preventive treatment of calcium spraying every 2 weeks in addition to ensuring regular irrigation of the plants, two measures that will prevent most of the damage. The use of geotextile is an additional method that we adopt to better conserve soil moisture, which is beneficial for this crop.

We pick the larger outer stalks to form bunches and secure them with rubber bands. We leave the heart intact, careful not to harvest the plants too severely, as this may reduce their future growth potential. A gradual harvest produces a more constant supply.

Dandelion

Cultivars

Italiko Red
Clio

Spacing

4 rows
6 inches (15 cm)

**Start in
indoor nursery**

30 days in a
72-cell tray

Transplant date

Mid-September

We grow dandelion in small quantities to diversify our winter production. To our surprise, it is very popular with our restaurant customers because of its distinctive flavor. Chefs are always looking for new tasty ingredients, and dandelion fits the bill perfectly. This member of the chicory family is particularly appreciated in Italian cuisine for its bitterness. Be sure to choose dandelion cultivars developed for their culinary taste. Our favorites are Clio and Italiko Red.

We grow it very densely in four rows, since its leaves do not cover much ground. We choose not to grow a lot, since it is not the most profitable vegetable because of the slow regrowth of its leaves in low-light periods.

Dandelion is fairly cold hardy but can suffer irreparable damage when night temperatures remain below 23°F (–5°C) for several consecutive days. For this reason, we prefer to grow it in our minimally heated greenhouse.

When the largest leaves reach 10 inches (25 cm) in length, we harvest it leaf by leaf in bunches of 12, taking care that several leaves stay on each plant to ensure a good regrowth.

Like sorrel, dandelion is a perennial plant that performs particularly well in early spring. If planted in the fall, it will be one of the first to provide abundant harvests in February and March.

Bok choy

Cultivars

Li Ren Choi
Mei Qing Choi

Spacing

3 rows
9 inches (23 cm)

**Start in
indoor nursery**

42 days in a
50-cell tray

Transplate date

September 2 to
early October

Bok choy is a fast-growing vegetable that is easy to produce. Although well-suited to colder temperatures, it can suffer irreparable damage when temperatures drop below roughly 7°F (−14°C).

Our favorite approach is the following: in early October, we transplant 3 rows of bok choy, intercropped with kale for bunching. Bok choy quickly reaches full maturity, while young kale plants grow more slowly. In December, we harvest the bok choy, leaving the kale all the space it needs to reach maturity. This strategy allows us to maximize the greenhouse space, which is a precious commodity.

When bok choy is grown alone, it's best to plan for a first transplant in September and another one in early January, in the same bed, to maximize available shelter space. For this second transplant, the number of days to maturity will be shorter due to increased sunlight each day after Christmas, which accelerates plant growth.

Chinese cabbage

Cultivar

Tokyo Bekana

Spacing

4 rows
6 inches (15 cm)

**Start in
indoor nursery**

42 days in a
50-cell tray

Transplant date

Early September to
mid October

We only grow one kind of Chinese cabbage in winter: Tokyo Bekana. It looks like lettuce, but it tastes like sweet cabbage. This cold-hardy variety continuously produces new leaves that we harvest in bunches throughout the winter. In our experience, its growth is one of the least affected by decreasing light.

As with most vegetables that we want to pick all winter, the planting date is critical and should be no later than mid-October. The size of harvests from December to March depends largely on the plant's ability to develop a strong root system before November.

Like Swiss chard for bunching, we prefer to cultivate Chinese cabbage in a minimally heated greenhouse to speed up plant growth for a continuous harvest. It is also possible to grow it in an unheated greenhouse with row covers as it can survive a light frost.

Tokyo Bekana cabbage can also be produced as a salad green. In this case, the spacing and climate guidelines for Asian greens apply. We cut the leaves above the growing tip, which allows for regrowth and several harvests. In mesclun, these leaves are delicious, adding a hint of pepper and a buttery texture that is unique to this cultivar.

Cilantro (coriander)

Cultivars

Calypso
Marino
Santo

Spacing

3 rows
9 inches (23 cm)

**Start in
indoor nursery**

42 days in a
72-cell tray
3 seeds per cell

Transplant date

Mid-September to
early October

Cilantro is a perfect herb for winter production. Since it can handle the cold, but does not tolerate much frost, we grow it in beds in our minimally heated greenhouse, to add variety to our winter offering. We intercrop cilantro with Swiss chard or kale for bunching, or we transplant it along the aisles of other crops.

To decrease the growing time in the greenhouse, we sow cilantro in our nursery 42 days before the scheduled transplant date. Once transplanted into the greenhouse, it can be picked in 30 to 40 days, leaving space for primary crops that we harvest continuously in winter. This technique optimizes the use of greenhouse space.

The cilantro harvest can be extended, to last a few weeks, by harvesting only a part of each plant in the bed.

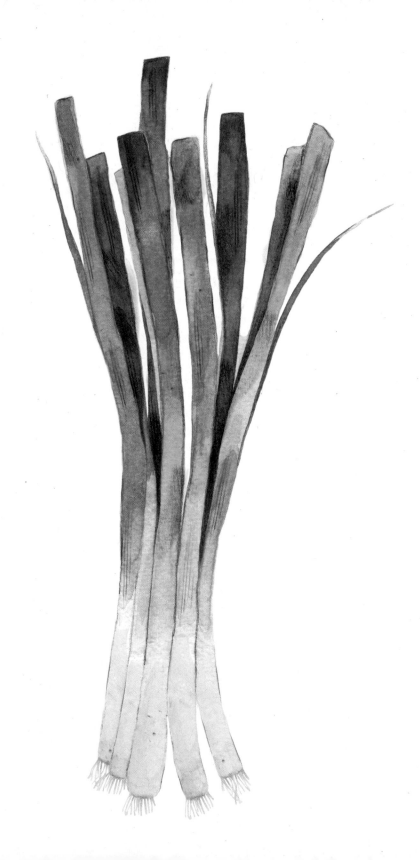

Green onions (Scallion)

Cultivar

Evergreen

Spacing

3 rows
9 inches (23 cm)

**Start in
indoor nursery**

55 days in a
72-cell tray
5 seeds per cell

Transplant date

Mid-August
to mid-February

We usually intercrop green onions with kale or Swiss chard. It's one of the reasons why this vegetable is often assigned to a minimally heated greenhouse where it will grow better than in an unheated one. Green onion is a worthwhile crop because it occupies very little space, grows relatively quickly, and can fill available bed space at the foot of other plants. This product is also a big hit with our customers.

When green onions are not intercropped, we transplant them in 3 rows in mid-August. Do not wait until after this date as their growth is considerably slowed by the decrease in fall light. Then, we harvest them for our winter holiday markets. This frees up space for multiple successions of quick-growing crops, such as arugula, baby kale, or baby mustard, which we seed in January.

To ensure the success of this crop, it's important to give your green onions a strong start in a nursery. We sow them into a 72-cell tray, 5 seeds per cell, and wait at least 55 days to transplant seedlings, which need to be at least 4 inches (10 cm) high.

In winter, our preferred green onion is Evergreen, the most cold-hardy cultivar available. In fact, it is so frost-resistant that it can even be over-wintered for a spring harvest. Seedlings transplanted into high tunnels in the fall only need one layer of row cover for protection. This allows them to grow all winter, until they are harvested early in spring, just in time for our first markets. If you opt for this approach, be careful not to delay the harvest too much as green onions may bolt once spring arrives.

Kale

Cultivars

Darkibor
Siberian
Winterbor

Spacing

2 rows
12 inches (30 cm)

**Start in
indoor nursery**

42 days in a
50-cell tray

Transplant date

Mid-August to early
October

Kale is one of our favorite winter crops. It is extremely simple to grow, easy to maintain, and very cold hardy, providing abundant harvests all winter long. Plus, when sheltered from the wind, the leaves become especially tender and flavorful. Our customers can't get enough of it.

Although kale grows quite well in a cold climate, we prefer to plant it inside our minimally heated greenhouse, which allows for faster regrowth and, as a result, greater yields. We like to intercrop it with vegetables that we harvest only once, like bok choy, green onion, or cilantro.

In mid-September, we transplant kale seedlings in 2 rows per bed; these were started 42 days prior to transplanting, in 50-cell trays. In the same bed, we transplant a second crop (like bok choy) in staggered rows. After 60 to 70 days, we harvest the entire bok choy plants, freeing up more space so the kale can reach its full size for the rest of the season. Because the crop stays in the same bed for many months, make sure to loosen up the soil with a broadfork before planting and, most importantly, fertilize the bed with a generous amount of compost. An additional nitrogen supply, such as a foliar spray, will also be needed when the plant is about two thirds of the way to maturity.

We start harvesting kale in December when the plants have several leaves and are well established. From then on, we harvest continuously, choosing the largest leaves to make small bunches. We make sure that a few sizable leaves are left on the plants to ensure better regrowth.

Komatsuna

Cultivars

Komatsuna

Spacing

4 rows
6 inches (15 cm)

**Start in
indoor nursery**

30 days in a
72-cell tray

Transplant date

Mid-September

Komatsuna is a leaf vegetable native to Japan. Also called mustard spinach, its leaves have a tender texture and very sweet taste. It's easy to cook but can also be eaten raw.

After we tried it, we adopted it! Its speed of regrowth is the main reason for that. Like bekana Chinese cabbage, komatsuna is one of the fastest-growing winter greens. Its seems to be not at all or very little affected by the seasonal light shortage, making it an ideal winter crop.

We grow Komatsuna very densely in an unheated greenhouse. Very cold hardy, it can endure temperatures below 14°F (–10°C) without suffering irreparable damage.

We harvest leaves in bunches of 10 to12 once they are at least 12 inches (30 cm) long, leaving the center to produce a continuous harvest. Since crucifers tend to go to seed in February, we aim to finish harvesting our komatsuna beds in early spring and then replace it with another crop.

Lettuce

Cultivars

Salanova
Green Incised
Green Sweet Crisp
Red Butter
Red Sweet Crisp

Spacing

5 rows
6 inches (15 cm)

**Start in
indoor nursery**

35 days in a
72-cell tray

Transplant date

Mid-August to
early October

We use Salanova leaves as the base for our mesclun. These cultivars were selected to develop only small leaves around the core, and lettuce heads can be harvested for mesclun at any stage of growth.

Most lettuce can survive a light freeze, if it lasts just a few days. In any case, this crop benefits significantly from a heated shelter. It increases the dollar-per-bed yield, and it also helps to reduce the incidence of fungal diseases like powdery mildew and downy mildew, which can be a significant problem in cold and damp greenhouses.

To get multiple harvests from each bed, you need to transplant the lettuce as soon as possible before early October. This will allow enough time to establish a strong root system before November. We can expect a productive crop to yield up to four cuts over one winter season. Harvests can be extended until May, then we clear the beds to make way for summer vegetables like beans and cucumbers.

That said, the success of winter lettuce is never a given. Pests, such as aphids, can quickly compromise an entire crop. To reduce the likelihood of an infestation, limiting nitrogen inputs when fertilizing is a proven strategy. From our experience, however, you should plan to run a phytosanitary treatment (biopesticide) at the first sign of an infestation. Some fungal diseases can also endanger Salanova, so good air circulation and moisture management can make all the difference. This crop also benefits significantly from frequent cultivation to decrease weed pressure.

We harvest Salanova lettuce by cutting leaves with a sharp knife, just above the growing tip. A clean cut is the best way to ensure a good regrowth for subsequent harvests.

Mesclun

Greens selection

Arugula
Baby kale
Baby Swiss chard
Claytonia
Mizuna
Mustard
Salanova
Sorrel
Spinach
Tatsoi

Mesclun is a mix of different small greens or salad greens. Because it is colorful, provides a variety of textures, and is sold prewashed, seasonal mesclun is one of our customers' favorite products.

We want to offer it year-round. The composition of our mixture is always evolving, depending on the availability of different greens over the seasons. In winter, our mesclun has remarkable flavor—a subtle taste, even becoming sweet—and takes on new textures. We grow several beds of crops (Salanova lettuce harvested as leaves, mustard, arugula, Claytonia, spinach, sorrel, tatsoi, mizuna, baby kale, and baby Swiss chard) to include in the mesclun or sell on their own. This approach to production provides tremendous flexibility.

Over the winter, as each crop reaches maturity, we integrate it into our mesclun without trying too hard to standardize the product. The idea is to sell a mixture that changes weekly and create a stunning combination of colors and flavors, based on availability. This method simplifies crop management, allowing us to roll with the punches thrown by the winter growing season.

The mesclun and the proportions for each ingredient come together in the packaging stage. To deliver an exceptionally clean product, we wash the leaves in a bubbler that also mixes the ingredients. The leaves are then dried and bagged. And our clients keep coming back for more!

New potatoes

Cultivars

Connect
Norland
Rose Gold

Spacing

2 rows
12 inches (30 cm)

Seeding date

Mid-March to early
April

Because potatoes are frost-sensitive (leaves can be destroyed at 28°F [–2°C]), we grow them in our minimally heated greenhouses, where temperatures never dip below 37°F (3°C). We begin the greensprouting, or presprouting, process in mid-February to sow potatoes in mid-March. This produces a first early harvest around mid-May, just as our summer markets are starting up. By choosing various cultivars with different days to maturity (DTM), we are able to extend the harvest period.

When planting, the soil must be relatively warm (at least 46°F [8°C]). This is why, one week before planting, we solarize our beds with a transparent tarp. Sufficient fertilizer is also a must. To make sure tubers will grow well and uniformly, it is is essential to loosen the soil with a broadfork before planting.

We maintain the crop by hilling once when the plants are 6 inches (15 cm) tall and again when they reach 12 inches (30 cm). This method suppresses most weed competition, encourages tuber growth, and keeps the crop healthy (as it could otherwise be susceptible to fungal diseases).

The first new potatoes are ready to harvest when the plants are flowering, 8 to 10 weeks after planting. We use a pitchfork and then gently wash them to conserve their delicate thin skin.

This is a highly profitable crop because it can be harvested early, producing small new potatoes that add variety to our offering. These factors easily justify growing this crop. Since we are among the first producers to bring new potatoes to market, they draw customers to our table, often building a relationship that lasts for the rest of the season.

Parsley

Cultivars

Peione
Giant of Italy

Spacing

3 rows
12 inches (30 cm)

**Start in
indoor nursery**

49 days in a
50-cell tray
3 seeds per cell

Transplant date

Mid-August to
September

As we looked to diversify our winter production, we explored which herbs were the most cold hardy. Our research and observations led us to identify parsley as one of the herbs with the most potential for winter production...and we weren't wrong! After growing it for a few years, parsley has become a must-have winter crop.

Parsley is one of the easiest crops to grow in a minimally heated greenhouse. With very little sensitivity to the fall light shortage, it grows really well in winter and seems unaffected by cold temperatures. The secret to success is very early transplanting in the greenhouse in the fall so that the crop matures before winter. We aim to have huge plants by November 1. Then, all we have to do is harvest it gradually during the following months.

To reduce the work associated with this crop, we produce parsley on geotextile in 3 rows spaced every 12 inches (30 cm). Drip lines under the geotextile irrigate the parsley as needed on sunny days.

The more we pick the parsley, the more it produces. For this reason, we harvest it every week, while making sure enough leaves remain to continue capturing the light necessary for growth. Our method is simple: we handpick the biggest leaves of each plant, one by one, and form bunches that we secure with a rubber band. This technique allows us to extend the harvest period of the parsley. If we cut the whole plant with a knife, it would be several weeks before we had another harvest. With the leaf-by-leaf method, we can pick continuously, which allows us to have more consistency in our offering.

Radish

Cultivar

French Breakfast

Spacing

6 rows
1.5 inches (3.8 cm)

Seeding date

Mid-September to
mid-February

Growing radishes in winter is challenging because, as the crop progresses into later stages of development, the root tends to grow long and thin under low-light conditions. Radish is, however, a highly popular vegetable in the spring, when demand is at an all-time high. Strategically, it's better to focus your efforts on growing stunning early spring radishes. In mid-February, we sow our first radishes in a high tunnel. Depending on the space available, we seed a few successions 10 days apart to ensure a continuous supply for our first markets in May.

As is the case with turnip, we sow radishes in 6 rows using a precision seeder. No thinning is required, only weeding with a flex tine weeder and wire hoe. Later, we harvest radishes in small bunches and sell them at a premium price.

Senposai

Cultivar

Senposai

Spacing

3 rows
12 inches (30 cm)

**Start in
indoor nursery**

30 days in a
72-cell tray

Transplant date

Mid-September

While attending a lecture given by farmer Pam Dawling, we were intrigued by senposai, an Asian green that she highly recommended trying because of its strong potential for winter growing. She has grown it successfully for several years in an unheated greenhouse on her farm in central Virginia.

Inspired by Pam's results and with no prior knowledge, we found senposai seeds and tested its winter production. In the first year, the crop was a success!

Senposai is a cross between komatsuna and regular cabbage. Its large rounded leaves are quite rigid and resemble collard. We harvest the leaves one by one and put five in a bundle, making sure the center of the plant is intact to ensure a good regrowth.

Senposai is appreciated by our customers as it is a very different vegetable from what we offer in winter. It can be used as a cabbage leaf for vegetable rolls, for example. However, the leaves regrow quite slowly, which does not make it the most profitable vegetable for winter production. Because of its cold tolerance and distinct character, it still deserves a place in our minimally heated greenhouse.

To leave enough room for its large leaves to develop, we grow senposai in 3 rows with 12 inches (30 cm) between each plant.

Sorrel

Cultivars

Red Veined
Sorrel (regular sorrel)

Spacing

5 rows
4 inches (10 cm)

**Start in
indoor nursery**

30 days in a
72-cell tray
2 or 3 seeds per cell

Transplant date

Early August to early
September

Sorrel is a low-maintenance crop that generates solid revenue. We grow it to improve our winter mesclun and to sell on its own. We appreciate the Red Veined cultivar for its leaves of contrasting colors from pale green to dark red. Chefs appreciate the plant's acidic flavor and never seem to grow tired of it.

Like several other greens that we want to harvest throughout the winter, we plant sorrel at the beginning of September so that it can get established well before November. Thirty days before the transplant date, we sow sorrel in our nursery, at two or three seeds per cell in 72-cell seeding trays. Sorrel is transplanted in 4 rows to maximize access to sunlight. Harvest this slow-growing crop when the leaves are the length of your palm. When deciding which crops to include in a winter production, sorrel is a shoo-in, due to its unique color and flavor.

We harvest the entire plant with a knife, cutting just above the growing tip, which allows the sorrel to regrow. Under the right conditions, we can expect to get three or four harvests over one winter season. In spring, plant growth accelerates, and sorrel will grow fast through March, April, and May.

Swiss chard (mature)

Cultivars

Fordhook Giant
Peppermint Rainbow

Spacing

2 rows
12 inches (30 cm)

Start in indoor nursery

42 days in a
50-cell tray

Transplant date

August 15 to
October 1

We grow Swiss chard for bunching in our minimally heated greenhouse. This allows us to harvest continuously over the winter, which is highly beneficial for us.

To grow a healthy crop, we transplant it into our greenhouse no later than October 1. This gives the plant almost enough time to reach maturity (plants are at least 12 inches [30 cm] tall) before day length begins to drop below 10 hours. Waiting any longer would be risky as Swiss chard requires a lot of light to establish a root system that will sustain robust regrowth and a continuous harvest. Choosing a cultivar known for its cold resistance is also an important factor to consider.

Because harvests will span several months, the crop must be well fertilized from the outset. When preparing the beds, we apply a generous amount of compost. An additional nitrogen supply will also be required once the plant is about two-thirds of the way to maturity.

Once the crop has reached maturity, we harvest a few outer leaves from each plant and bunch it every week. The best way to make sure that new leaves will regrow quickly, especially in low-light conditions, is for some mature leaves and the heart of the plant to stay intact. With this strategy, we can harvest the same Swiss chard plants until March. When the time comes to make room for summer crops, we harvest the entire plant and sell it as-is.

Turnip

Cultivars

Hakurei

Spacing

6 rows
1.5 inches (3.8 cm)

Seeding date

Mid-September or
mid-February

Transplant date

August 15 to
October 1

Sometimes called salad turnip or just Hakurei, this small white vegetable hails from Japan. We are crazy about it, but it is tricky to grow in winter conditions. In a cold climate, turnip becomes quite sweet and tender. It's no surprise that turnip is a crowd favorite among our clients…and among small rodents that can, in a single night, take a bite from each and every root, thus spoiling the crop.

For a successful turnip crop, we recommend sowing at the very start of fall for a harvest before Christmas, and again in February for an early harvest in May. These seeding dates allow us to produce a high-quality root vegetable. When, instead, turnips are overwintered (i.e., sown later in the fall and harvested in February and March), the root tends to grow long and thin. Around mid-March, these turnips go to seed, rendering them worthless.

Turnip leaves are frost-sensitive and require minimal heating or protection from a row cover to keep temperatures above freezing. Rodents must be systematically trapped to stop them enjoying our turnips.

We sow turnips in 6 rows, using a precision seeder, then sell fresh harvests as bunches with leaves still attached.

Watercress

Cultivars

Upland Cress
Watercress

Spacing

4 rows
1.5 inches (3.8 cm)

Seeding date

Mid-August to early
October

Watercress adds a bright flavor that is an excellent complement to lettuce leaves and other mesclun greens. Most importantly, watercress is heavy, which delivers a worthwhile dollar-per-bed ratio. It's worth mentioning, however, that it is slow to become established, especially in winter conditions when there is little sunlight. Watercress needs to get an early start in the fall so we give it a spot in our minimally heated greenhouse. Unlike other winter greens, it is only somewhat cold hardy.

Once planted, it becomes quite dense, and for this reason, we only seed it in four rows. It can be harvested two or three times in a winter season. To extend harvests until early spring, choose the Upland Cress cultivar because it is less likely to bolt than Watercress. Whichever you choose will need constant drip irrigation for healthy growth. This is why we plant it next to arugula, Chinese cabbage, and other vegetables that benefit from a constant water supply.

Asian greens

Cultivars

Koji
Mizuna
Tatsoi

Spacing

6 rows
0.5 inches (133 cm)

Seeding date

Mid-September to
mid October

Tatsoi and Mizuna are two species of greens that we grow in winter to add to our mesclun. These are both safe bets, but there are so many different varieties of Asian greens that it's worth experimenting with new ones. Most of these are cold hardy and can be grown in an unheated greenhouse if insulated by a row cover.

We seed them in 6 rows and harvest them two or three times in a winter. Tatsoi, which takes longer to grow than Mizuna, must be harvested more carefully to ensure good regrowth. It will also be quick to bolt if greenhouse temperatures are too high, so keep an eye out for this. To be safe, plan to harvest Tatsoi several times before February.

Baby kale

Cultivars

Red Russian
Siberian

Spacing

6 rows
1.5 inches (1.3 cm)

Seeding date

Mid-September to
early October

When combined with mustard, tatsoi, and Swiss chard leaves, baby kale is an essential ingredient in our winter mesclun. Very well suited to cold climates, this crop is one that we can rely on. Red Russian leaves are particularly attractive, with purple stems, and have a milder flavor when grown in cool temperatures.

We plant baby kale in 6 rows and manage to keep the crop weed-free by regularly running a flex tine weeder over the bed. After harvesting it with a knife, we take the time to rake the bed, removing any debris that was left behind. This ensures that the crop will have healthy regrowth, ready for a second cut.

Baby Swiss chard

Cultivars

Bright Lights

Spacing

6 rows
0.5 inches (1.3 cm)

Seeding date

Early to
Mid-September

Baby Swiss chard is one of our favorite winter crops. Cold weather causes the stems to develop bright colors, which makes this a unique ingredient in our winter mesclun. Our rainbow cultivar also enhances this effect, with different-colored stems that range from yellow and pink to orange and red.

We plant 6 rows of Swiss chard per bed, with an in-row spacing of 0.5 inches (1.3 cm). Because this crop is planted in early September, we harvest each bed several times over the winter, even in an unheated greenhouse. To get multiple harvests, we make sure to cut the leaves above their growing tip.

Claytonia

Cultivar

Claytonia

Spacing

4 rows
1.5 inches (3.8 cm)

Seeding date

Mid-September

Claytonia, also known as miner's lettuce, has small light green, pointed leaves and a very mild taste. It is the most cold hardy of our greens and is sold on its own and in mesclun.

We like to maintain several Claytonia beds for harvesting all winter long. Because it is so capable of handling cold weather, we often plant it at the ends of our greenhouses, in the beds that are most exposed to outside temperatures.

We seed Claytonia in four rows, and it becomes quite dense at maturity. From each bed, we get two or three harvests per season. In early spring, Claytonia tends to bolt, so we pull it out of our tunnels around mid-March to make room for a summer crop.

Early carrots

Cultivars

Bolero
Mokum
Napoli

Spacing

4 rows
1.5 inches (3.8 cm)

Seeding date

Mid-February to late
march

In summer, we sow carrots every week to ensure a continuous supply. Our customers expect to be able to buy carrots that are exceptionally fresh. Our last seeding is a storage carrot that we harvest and keep in our cold room all winter to sell on demand. This classic method is the most efficient approach to providing carrots in the winter season.

Storage carrots are so popular that, despite all our planning, we are unable to grow enough to satisfy the demand. Usually, our supply runs out in February. As a result, the concept of growing fresh carrots in winter months, to offer early spring harvests, is quite appealing. They do well in cold climates, starting in February, when there is enough sunlight to sustain growth. Our first early carrots are sown in a minimally heated greenhouse, and the next successions are seeded in our high tunnels.

We seed carrots in 4 rows with 1.5-inch (3.8 cm) spaces between rows, to optimize access to light and thus promote rapid growth. To increase soil temperature in high tunnels, we sometimes use clear plastic tarps to solarize beds before seeding. Our favorite variety is Napoli, but we have also been experimenting with Mokum for the past few years, and we are very fond of it.

Quick tip for selling early carrots in the spring: the selling price needs to be twice the summer rate. This can easily be justified by the slow growth and, especially, the popularity of this crop.

Mâche (corn salad)

Cultivar

Vit

Spacing

4 rows
1.5 inches (3.8 cm)

Seeding date

Mid-September

Mâche is a unique green that has a velvety texture, with a sweet and pleasant flavor. This very cold-hardy vegetable is well-known among our French customers. In Europe, mâche is the ultimate northern crop, and many people here have told us how much they enjoyed discovering it. Because it isn't as well-known in Quebec, mâche is an object of curiosity among our customers.

Unfortunately, this crop is not particularly profitable, due to poor regrowth after the first cut, so we harvest the entire rosette in a single cut. Mâche reaches maturity in 60 days when sown in mid-September. We seed mâche densely, in four rows, using a precision seeder.
Once mature, it will keep well, which means harvests can be extended over several weeks, until March. We cut this crop with a knife, picking the rosettes close to the ground.

Later seedings in the fall are also possible for plants that will overwinter in a high tunnel for harvesting in early spring. However, without minimal heating, there's no guarantee that this crop will succeed. Either way, plan to harvest the last round of mâche before May as it does not tolerate heat and will go to seed.

131

Mustard

Cultivar

Garnet Giant
Ruby Steaks
Scarlet Frills

Spacing

6 rows
0.5 inches (1.3 cm)

Seeding date

Mid-September to
mid-October

Mustard, one of the base ingredients in our winter mesclun, grows fast and has a strong frost resistance. Unlike mustard grown in the summer, winter mustard is quite mild. We love Ruby Streaks and Scarlet Frills, which have ruffled red leaves that can be harvested at any size.

We sow mustard in unheated greenhouses and protect it with row covers when temperatures drop below freezing. This strategy is enough to keep the crop alive, provided the row covers are supported by hoops. They prevent contact between the mustard leaves and the fabric, which would cause considerable damage during a freeze event.

We grow it in 6 rows, as opposed to our 10-row summer seedings which provides better light penetration and air circulation in the crop canopy. Mustard will yield three more cuts after the first harvest. It can last the entire winter season since we harvest only a section of mature beds every week.

Spinach

When it comes to winter vegetable production, spinach is the star of the show. We grow as much as we can and still struggle to meet our customers' demand. Spinach is perfectly suited to winter conditions. It is at its best in the cold season when it develops a delicious, sweet flavor and a perfect texture.

After several rounds of trial and error, we now grow spinach exclusively from transplants. This allows us to plant it quite late in our high tunnels, which are taken up by our summer vegetables, like peppers and Italian tomatoes, until at least mid-September.

While transplanting spinach does require more work than direct seeding, it ensures better uniformity and therefore improved yields. Because the crop will be harvested all winter, this additional step is well worth the time investment.

We harvest spinach every week, alternating between beds. For a continuous harvest, we cut only the outer leaves, individually; the remaining five to seven leaves are enough to continue sustaining rapid growth.

Since spinach is quite cold hardy, we grow it in our unheated high tunnels. We recommend using row covers on freezing-cold days and nights, then removing them on sunny days to increase access to light and promote photosynthesis.

When space is available, we also grow spinach in a minimally heated greenhouse. This strategy is particularly profitable in the spring when demand is at an all-time high.

Tools for
Winter Growing

❄

In short, the market-gardener hones his craft, continually striving to improve his growing practices to increase his income; he is not afraid to make significant amendments to the soil, so that he may take from it everything that it can provide. And he works with his own unique tools.

— Joseph Vercier, *Culture potagère*, 1911

Greenhouses, tunnels, and other shelters used to grow vegetables in winter are tight spaces. For highly mechanized farms, the change of season is problematic because tractors are not designed to be used in shelters. For market gardeners, who operate on a human scale, it's easier to pivot towards winter farming as we see it.

As we have shown, growing vegetables in winter is no simple task, and it comes with its fair share of challenges. Several factors may complicate the endeavor, but fortunately there are solutions available to counter the effects of limited sunlight, cold, weeds, slow plant growth, etc. Among these solutions are a range of tools that modern-day growers can use to ensure the success of their winter production. We get the most out of these tools by paying careful attention to each task.

The Essentials

THE BROADFORK

The broadfork is an indispensable tool for non-mechanized soil preparation. A large U-shaped fork with two handles, it is generally 24 inches (60 cm) wide, with four to six steel tines that are roughly 12 inches (30 cm) long. We use it to prepare beds before sowing or transplanting a crop.

Since the broadfork doesn't turn over soil layers or generate compaction, it aerates the soil without damaging its structure. Good soil preparation is one of the most effective strategies to ensure the success of a crop; and loosening the ground with a broadfork is the first step in this process.

When using a broadfork, we hold one handle in each hand and set the tines, with the tips pointing away from us, on the soil surface. We step on the horizontal bar, making the tines sink into the ground under our body weight. The effort required in this stage will depend on the soil type, the level of compaction, and how rocky the ground may be. Once the teeth have sunk into the ground, we step back and pull the handles towards us until they make a 45-degree angle with the soil surface. This angle is enough to open up and aerate the soil. We slide the broadfork tines along the bed for about 12 inches (30 cm), then repeat the process.

Once we've loosened an entire bed with the broadfork, we fertilize the soil with compost, pelleted chicken manure, etc. To integrate the amendments, we use a tool like a power harrow, a wheel hoe, or a Tilther.

THE BED PREPARATION RAKE

A good bed preparation rake is an indispensable tool for any market garden. Distinguishing features are a flat head and angled rigid tines spaced 4 inches (10 cm) apart. Opt for a rake that is 30 inches (75 cm) wide, so you can work the full width of a bed in a single pass. We use this versatile tool almost daily to accomplish different tasks, especially preparing permanent beds before planting a new crop.

When it's time to transition to winter crops in our shelters, we use the rake to clear the bed surface of plant residue built up over the summer. We then use it a second time to evenly distribute amendments on the bed. We also use the flat back of the rake to spread compost, ensuring that we've created a uniform soil surface.

Once the beds are prepped and ready for the next seeding, the secret is to rake them once again. This will carefully deposit excess rocks and debris into the aisles, which we then gather and remove from the shelter. When it comes to direct seeding, a uniform surface provides a better likelihood of success. It allows the seeder to roll unimpeded along the bed, distributing seeds onto the soil without a hitch.

The bed preparation rake can also be used to mark rows on the bed surface before transplanting a crop. To do this, we use row markers or cross-linked polyethylene (PEX) tubing. We slide the tubes onto the tines that match the crop's row spacing, then pull the rake down the length of the prepared bed. At this stage, we make sure to walk in a straight line as this will facilitate weeding. When marking rows with this technique, we create a grid by running the rake along both the length and width of the bed. Where the lines cross, a seedling will need to be transplanted.

Bed Preparation

THE TILTHER

The Tilther is a small rototiller that runs on a battery-powered drill. It is guided using two wooden handles. Designed by Eliot Coleman, this tool is perfect for working soil in a tight space like a greenhouse or tunnel. For vegetable farms that are starting up and for home gardens, the Tilther offers a good alternative to the walk-behind tractor as it is affordable and easy to use.

It's also the ideal tool to quickly prepare a bed between two crop successions, or to efficiently integrate amendments like compost.

The Tilther works on the same principles as a traditional rototiller: it has teeth with a 90-degree bend, which rotate along a horizontal axis. While repeatedly turning over the soil will generally compromise its structure, the Tilther only penetrates the ground to 2 inches (5 cm). This leaves the soil structure intact, so that microorganisms can continue to thrive. In fact, the Tilther is considered to be a tool that aligns with minimal tillage practices.

The Tilther we use is especially helpful in shelters, but we don't recommend it for rocky or compacted soils, or in heavy soil, such as clay.

THE WHEEL HOE (GLASER)

If you don't have a walk-behind tractor or Tilther, you can prepare beds with an oscillating wheel hoe (often called a Glaser wheel hoe or just wheel hoe). This two-handled tool is the simplest and most economical method to till soil. It has two handles that, with a good deal of energy, are used to push and pull the hoe.

To prepare a bed with a wheel hoe, simply set the oscillating blade down on the bed after spreading fertilizer onto the surface. Then, move forward while pushing and pulling on the two handles, making the blade cut through the soil to a depth of 0.75 to 1.5 inches (2 to 4 cm). The inflatable wheel at the front reduces the amount of effort required, while the back-and-forth motion mixes the fertilizer into the soil. There's no need to push the blade deep into the ground as this requires excessive effort and damages the soil structure. A bed is ready for transplanting or direct seeding once the blade has cultivated the surface.

With a 12-inch (30 cm) blade, the wheel hoe is also ideal for weeding aisles in shelters. In the winter, we cultivate the aisles about once a month, or when the weeds have reached the cotyledon or white-thread stage. You can also use the wheel hoe to quickly uproot a crop before removing it from the bed and transplanting a new crop—all in the same day. In winter, this method is particularly useful as we want to grow something in every bed at all times to maximize our use of space.

THE WALK-BEHIND TRACTOR

The walk-behind tractor consists of an engine mounted on two wheels and equipped with a power take-off (PTO). Accessories can be hooked up to carry out different tasks, especially to work the soil. We steer the device using two handles, as we walk down the aisles. Professional-grade walk-behind tractors are sturdy enough to carry out tasks that include shredding cover crops, preparing soil, and shaping permanent beds. At Ferme des Quatre-Temps and in many small-scale market gardens, the walk-behind tractor can eliminate the need for a standard four-wheel tractor. We recommend the BCS 853 model, which is ideal for connecting 30-inch (75 cm) wide accessories that span the full width of a bed.

In winter, we generally use a walk-behind tractor to prepare beds in shelters. This tool speeds up the process and makes it much easier. We choose to primarily work with the power harrow because it incorporates amendments effectively and loosens the soil without disturbing it unnecessarily.

THE POWER HARROW

The power harrow is a walk-behind tractor accessory, primarily used to prepare the soil before transplanting or seeding. It has several metal blades that rotate on a horizontal axis. The most interesting feature of this harrow is that the blades mix the soil like an eggbeater, but they only penetrate the top few inches of soil, without turning over layers. As a result, this preserves the soil structure and limits both compaction and the likelihood of developing a hardpan.

Behind the power harrow, a roller flattens the soil evenly, creating an optimal surface for seeding. We use this tool after loosening the bed with a broadfork and spreading fertilizer on the surface. Once the power harrow is attached to the tractor power take off (PTO), we adjust the control handles, turn on the PTO, put the tractor in first gear, and begin moving forward along the bed while walking down the aisle. After running the harrow down the bed, we sow or transplant a new crop.

Soil work done with a power harrow can also be done with a rototiller hooked up to a walk-behind tractor. That said, the rototiller will turn over soil layers (horizons), break down the soil structure unnecessarily, and till at an excessive depth. Because of this negative impact on soil health, we do not recommend using a rototiller in your shelters.

THE ROTARY PLOW

The rotary plow can also be hooked up to a walk-behind tractor. It consists of four 10-inch (25 cm) wide metal blades that form a spiral, like an infinite screw, that work down to a depth of 12 inches (30 cm). A wheel at the back of the rotary plow allows users to adjust blade depth.

We use the rotary plow to form our permanent beds. Unlike the power harrow, which runs along the bed surface, we run the rotary plow down the aisles. As the plow's spiral turns, it pulls soil from the aisle and moves it onto the adjacent bed. This displaced soil helps raise the bed surface. We could achieve the same result with a shovel, digging soil out of the aisles and onto the bed, but this would be highly time-consuming and laborious. For a professional market gardener, the rotary plow is a must.

Before planting winter crops in a shelter, it's best to use a rotary plow to reform raised beds, which have better drainage and generally warm up faster. This represents a significant advantage in winter, when soils tend to retain excessive moisture. Note that to create uniform beds and straight aisles, the rotary plow must be run in both directions down every aisle. We repeat this operation on each side of the bed.

148

148

Seeders

THE SIX-ROW SEEDER

The Six-Row Seeder is one of several precision seeders that are well-suited for work on small-scale farms. It is highly effective for sowing small to medium-sized seeds. The seeder features two rollers: a front roller flattens and levels the ground, while the back one drives a metal shaft, dropping and burying the seeds.

As needed, the seeder's 60-inch (1.5 m) handle can be offset to the right or left. This angled handle allows users to walk in the aisle while seeding in a raised bed.

The Six-Row Seeder is equipped with six containers, called hoppers, that distribute seeds. Because the seeder is 15 inches (38 cm) wide, it can cover half the bed in one pass. This is why it is often used to sow 12 rows per bed in one round trip. With this technique, we can set a 2.25-inch (5 cm) spacing between each row.

In winter, the Six-Row Seeder can be used to sow greens (arugula, tatsoi, mustard, baby kale, etc.) in shelters. To seed them in six rows, according to our winter spacing, we leave every second hopper empty, leaving 4.5 inches (10 cm) between rows. To ensure a successful seeding, the beds must be carefully prepared because this implement is highly sensitive to soil irregularities (debris, rocks, etc.). We want a perfectly level, refined soil surface. For a meticulous soil preparation, we first run the seeder down the bed with empty hoppers. If the seeder jams at this stage, we use the bed preparation rake on the surface again to remedy the situation.

Once the soil is ready, we set the seeder to the right seeding density by adjusting the pulleys and finding the required hole size along the metal axis. We also adjust the brushes in the hoppers according to the rate at which we want the seeds to drop in the row. After placing an equal amount of seed in each hopper, we put the seeder on one side of the bed and push it forward while walking along the aisle. On the return trip, we set the seeder on the other side of the bed to complete the seeding. Throughout the process, we keep an eye on the hoppers: if one is draining faster than the others, the brush might need to be lowered. It is helpful to weigh the seeds before and after sowing to find out how much actually dropped into the soil. Then, we make changes accordingly for our next seeding, or run over the bed once more to get the right density.

THE JANG SEEDER

Of all the seeders, the Jang is the most accurate and easiest to use. It is our favorite and an essential tool for any small-scale professional farm. Depending on the model you select, it will be equipped with one or more transparent hoppers.

To sow a crop, you need to adjust the gears to match the seed you've selected, then insert the correct seed roller into the hoppers. With these adjustments, seeds drop onto the bed at a constant rate. Next, weigh the seeds, fill the hoppers evenly, and mount them onto the seeder. Now all you have to do is push the seeder along the bed at a constant speed. In one round trip, the crop will be seeded throughout the bed.

Once you've tried the Jang, there's no turning back. Its greatest strength is its ability to achieve a dense and uniform direct seeding, even on an irregular surface (residue, rocks, etc.). For this reason, it provides a significant advantage when compared with the Six-Row Seeder. The Jang is also a major time-saver, seeding several beds in a row without the user having to constantly stop and fix the surface (removing debris, smoothing the ground, etc.).

With good bed preparation, the proper Jang adjustments, and a reliable seeding chart, even an inexperienced person can successfully sow a crop.

THE EARTHWAY SEEDER

The Earthway is the most affordable seeder on the market. Although able to sow any vegetable, it is most effective with larger seeds. Its mechanism is remarkably simple: the seeder creates a furrow in the soil, drops a seed, then covers the furrow with a chain dragging along the ground. As the front wheel turns, it allows seeds to drop onto the soil.

For all operations, from a large home garden to a small farm, this tool is the ideal first seeder. As soon as your garden begins to expand, however, the Jang or the Six-Row Seeder will become an essential asset because of its accuracy and capacity to sow multiple rows at once.

To use the Earthway, first prepare the bed surface and choose the appropriate seed plate, according to seed size. Then, adjust seeding depth and set up the row marker (if needed). The last step is to weigh the seeds and push the seeder down the bed.

FOUR-ROW PINPOINT SEEDER

With a very simple design, the Pinpoint Seeder can replace the Six-Row Seeder. It has four hoppers set close together and supported by a wooden handle, allowing users to sow 12 rows of greens in one bed. It is more affordable and more user-friendly than the Six-Row Seeder, but it requires three passes, rather than two, to complete the same seeding in a bed.

Before using the Pinpoint, adjust the seed shaft to match your seed size, then weigh and pour the seeds into the four hoppers. To start sowing a crop, pull the seeder towards you (this is the opposite of other models, which are push activated). Because this seeder doesn't have a roller to cover the seeds after dropping them, we recommend going over them with a flex tine weeder, to promote a better germination rate.

Hoes and Weed Control

THE STIRRUP HOE

The stirrup hoe is an essential weeding tool in any garden. It consists of a rectangular blade mounted on a long wooden handle. The blade oscillates when moved back and forth along the soil surface. Several blade widths are available (3 ¼ in., 5 in., 7 in. [8 cm, 13 cm, 18 cm]) for different weeding requirements. With a highly ergonomic design, this tool allows users to weed while standing, which speeds up the work and makes it more comfortable.

When using a stirrup hoe, we match the blade width to the distance between the rows to be weeded. The goal is to cultivate the entire soil surface, so we avoid using any blade that would be too narrow to span this gap between rows. We start by setting the blade on the ground, then push it back and forth, creating an oscillating motion. At this stage, there's no point in applying significant pressure and driving the tool into the ground. This would require more effort and would not improve efficiency in weeding.

153

A working depth of ⅓ inch (1 cm) below the soil surface is sufficient. The stirrup hoe is designed to separate the above-soil component of the plant from its roots. It functions best when used to kill weeds in the white-thread or cotyledon stage. Deprived of its reserves, the young plant will die. The stirrup hoe can also kill weeds in more advanced stages, but with a lower rate of success.

To get the most out of your stirrup hoe, you must sharpen the blade at least once a month. Another good tip: avoid using this tool on beds irrigated with a drip system because there is a high likelihood that the sharp blade will damage the plastic and cause leaks.

THE COLLINEAR HOE

The collinear hoe is a cultivation tool featuring a wooden handle and a fixed steel blade that is 7 inches (18 cm) wide. This tool's unique attribute is a 70-degree angle between the blade and the handle. As with the stirrup hoe, the collinear hoe can be used comfortably while standing with a straight back. The angle of the blade makes for an even more ergonomic design. For optimal comfort while weeding, hold the handle with thumbs pointed up.

The position and shape of the blade distinguish this model from other hoes. With the collinear hoe, you can reach weeds developing below the canopy of crops growing close to the ground, like cabbage and lettuce.

Simply drag the blade along the ground, keeping it roughly ⅓ inch (1 cm) below the soil surface. Make sure to disturb the entire bed surface, including near vegetables and below low-lying foliage, but be careful not to accidentally sever the roots of the main crop. This hoe can kill weeds at a more advanced stage (true leaf), though it is always better to weed sooner.

THE WIRE WEEDER

The wire weeder is a cultivating tool comprising a handle and an interchangeable tip. Each tip consists of a wire bent to suit a particular task. Unlike most cultivation tools, this one does not have a sharp blade, making it safe to use in drip-irrigated crops, without fear of creating leaks or damaging equipment. It provides efficient weed control by uprooting weeds entirely.

To use the wire weeder, simply install the tip that is most suitable for your task and start cultivating the bed. The tool only needs to reach a depth of about ⅓ inch (1 cm) to weed effectively and without too much effort. There's no need to work the soil any deeper. As with most cultivation tools, this one performs optimally when weeds are quite small (white-thread or cotyledon stage). Once they're beyond this stage, the best approach is to use the stirrup hoe, which can sever bigger weeds.

BIO-DISCS AND THE TERRATECK WHEEL HOE

Bio-discs are weeding tools installed on a Terrateck wheel hoe. They are equipped with two pairs of discs: one with small flat discs (on the inside) and one with large parabolic discs (on the outside). The wheel hoe acts as a support structure for these discs. Like the oscillating wheel hoe (Glaser), this two-wheeled metal structure is steered with two handles.

We use Bio-discs to weed most of our transplanted crops (cilantro, green onion, etc.) and some direct-seeded crops (carrot, baby chard, etc.). The main advantage with this tool is that it can cultivate vegetables on the row, meaning between each plant along the bed. Plus, it creates a gentle hilling effect at the base of the plants.

We start using Bio-discs when transplanted seedlings have taken root in the soil or as soon as the rows of a direct seeding are well-established. First, we set the hoe over a crop and move the discs along the notched metal bar to adjust the width between discs. When there are more rows on a bed, the discs must be set closer together. To weed a bed, we push the wheel hoe, keeping the crop safely located between the discs. This step is easier when rows are straighter. At this stage, make sure that the discs aren't burying or uprooting the crop. If they are, increase the distance between the discs. We also recommend removing the inner stationary discs for rocky soils, which will make it easier to use the tool.

To weed an entire bed, push the Bio-discs along every row, which results in a hilled crop with an aerated soil surface. Bio-discs save us time by allowing us to accomplish, in one fell swoop, all these tasks that would normally require several different tools.

FLEX TINE WEEDER

The flex tine weeder is a rake used for weed control. With dual rows of flexible tines, the flex tine weeder cleverly imitates tools used in larger-scale farming. Thanks to flexible metal tines, it can destroy weeds while—as if by magic—leaving the main crop intact. The secret behind flex tine weeding lies in the fact that the weeds are smaller than the well-established crop.

This is an essential tool for growing greens (arugula, baby kale, mustard) and radishes since it eliminates the need for hand weeding. We recommend the model with a 30-inch (75 cm) width as it cultivates the entire bed surface in a single pass.

It's important to note that this tool only works with very young weeds (from white-thread to cotyledon stages). Pay close attention to crops and make sure to use the flex tine weeder when the time is right, usually one week after transplanting. For direct-seeded vegetables, we wait until they reach the true leaf stage.

First and foremost, we choose a sunny day and make sure the soil is relatively dry. These conditions will help increase weed mortality rates. Next, we simply set the flexible tines in the middle of the bed and walk briskly along the aisle, keeping the tool slightly behind us. Depending on how well-established the crop's root system is, you may be able to exert greater pressure on the tool. This creates a more powerful weeding action, so keep an eye on the crop to check it isn't being damaged. In our experience, it's best to go over each bed twice, in opposite directions, to ensure most weeds have been pulled.

For direct-seeded crops, use the flex tine weeder about once a week, depending on the season. In winter, when there is less sunlight, weed growth slows, and we use the tool once a week or every two weeks. We repeat this process until the crop is well-established and capable of competing with weeds.

Harvest Tools

THE HARVEST KNIFE

When it comes to harvesting winter greens, such as arugula and mustard, we recommend using a sharp knife for the first cuts. Our favorite is a pocketknife made by Opinel, a classic French company. With a sharp knife, we can be sure to harvest every leaf at the right height, above the plant's growing tip. This tip is usually at the center of the plant, where the leaves meet. By cutting each leaf at the right height, we can ensure a second, a third, and sometimes even a fourth high-quality harvest. For winter crops that are planted once and harvested over several months, paying this extra attention while harvesting is well worth the investment. At this stage, we also cut the leaves to a maximum length of 4 inches (10 cm).

To quickly harvest greens with a knife, we grab a handful of leaves in one hand and cut them with the other. Diligent weed control will significantly speed up the process, allowing for a continuous harvest without needing to sort leaves.

THE QUICK-CUT GREENS HARVESTER

The greens harvester is an essential tool for market gardeners looking to produce a large amount of mesclun or greens (baby kale, mustard, arugula). It consists of a drill that powers a sharp blade and a metal shaft equipped with several green macramé brushes. The brushing motion is like an automatic car wash, with ropes spinning along the axis. These push freshly cut leaves towards a fabric basket, where they are collected before being transferred into a harvest bin.

This tool is an incredible time-saver compared to harvesting with a knife. However, it delivers a less precise cut and an inconsistent regrowth quality. For this reason, we rarely use it in our winter crops, but it can be a good option for the last cut of the season in a given bed. Since regrowth is no longer important at this stage, it's worth pulling the greens harvester out of hibernation.

To use it, we first install the drill on the designated support arm and set up several harvest bins along the bed. We turn on the drill to begin harvesting, carefully setting the blade height according to the required distance from the ground and optimal leaf length. When the blade is in motion, we keep our legs far out of the way. The blade is extremely sharp, and accidents can happen in a flash. We move slowly, while slightly lifting the tool at regular intervals to help the leaves slide into the bag. This looks like a rolling motion. Once the bag contains four to five pounds of greens, we empty it into a bin and continue harvesting. After the last harvest, we immediately uproot crop residues with a wheel hoe, then remove them with a bed preparation rake, to ready the bed surface for the next succession.

Tending to
Winter Crops

❄

All that is to last
is slow to grow.

— Louis de Bonald

In winter, all rhythms shift. Sunlight diminishes, days become cold, and nights are frigid. Plant growth slows significantly. To produce vegetables in winter, we must fall in step with the season's tempo, far removed from the cadence of summer. Farming methods also must change accordingly. This means adjusting fertilization and irrigation plans, as well as redefining crop maintenance and protection to account for the specificities of winter.

Because they grow at a slower rate, winter vegetables provide lower weekly yields than their summer counterparts. Growers must accept this reality and adopt a kind of serenity in the face of this unhurried pace. The key to a profitable winter market garden is limiting, as much as possible, the hours of labor required to maintain crops. Growers have only one option: to match the rhythm set by the plants. It's also important to remember that market gardeners need to find some rest in this season. Running a diversified vegetable operation in the summer is nothing short of a marathon. While growing in winter comes with its own share of challenges, they don't typically require urgent attention. Nature takes its time, and so too must humans. Market gardeners have to regain their strength so that, come spring, they'll be able to shift gears and accomplish the myriad tasks paving the way for an abundant summer season. This is a crucial consideration, and we want to remind everyone who may be interested in winter farming. Neglecting the human aspect of production will lead to significant long-term health risks.

That said, while the winter growing season requires a more relaxed approach, daily attention in greenhouses and tunnels is paramount. We have already seen the relevance of using the right tools for each task, but it is just as important to be familiar with best practices as a whole, to optimize all stages of production. Starting in September, growers need to switch off cruise control, step away from the torrid pace of summer operations, and move into winter mode. Here is an overview of our methods at Ferme des Quatre-Temps.

Challenge of Irrigating in Winter

In winter, plant growth slows, water uptake tapers off, and the work related to crop irrigation winds down. That's the good news. The bad news is that irrigation in very cold weather is a real challenge in unheated greenhouses. Water freezes in conduits, which can lead to significant damage (e.g., valves bursting or irrigation lines freezing). To successfully supply water to unheated shelters, the easiest solution is to install an antifreeze hydrant inside the greenhouse (also known as a frost-free hydrant or anti-siphon faucet). This faucet needs to be connected to a water line buried at a depth of 5 feet (1.5 m), lying below the frost line. When we open the valve, water rises in the pipe and we can irrigate crops despite below-freezing temperatures. When we close the valve, the water drops back down into the pipe, below the frost line. This freeze-proofing feature keeps water from sitting in the valve, which is the main cause of frost damage in an irrigation system. Installing this valve requires excavation work, but it is essential. When irrigating an unheated shelter without an antifreeze hydrant, valves must constantly be purged—and the threat of damage is always looming.

Anti-Freeze Hydrant

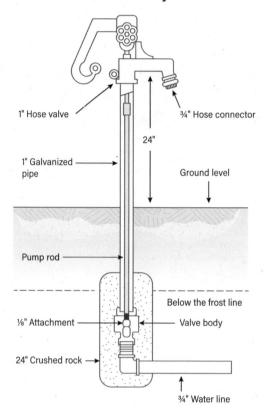

1" Hose valve

¾" Hose connector

24"

1" Galvanized pipe

Ground level

Pump rod

Below the frost line

⅛" Attachment

Valve body

24" Crushed rock

¾" Water line

The type of irrigation system you select, either drip or sprinkler, needs to take into account the likelihood of freezing temperatures. If the shelter is unheated, drip irrigation is the best choice. Make sure to install connectors properly to prevent water from pooling and stagnating. Irrigation lines in a drip system can handle some expansion in below-freezing temperatures and are less likely to break. Easy to maintain, the drip system is a highly effective solution for many vegetables. It also has the advantage of delivering water right onto the ground, keeping the plant foliage dry. This is particularly relevant in winter, when relative humidity tends to be quite high.

For direct-seeded crops sown in multiple beds at once (still in an unheated shelter), drip is not an optimal irrigation system because it's impossible to lay out the lines before the seeds germinate. In this case, after the seeding, sprinklers must be set up to water the beds until the crop germinates and becomes established. To avoid any frost damage, irrigation lines must be purged and/or stored in a heated place, such as a garage, after being used.

In minimally heated greenhouses, the threat of freezing temperatures is nonexistent and equipment damage is less likely. In this case, choosing between drip or sprinkler irrigation will depend instead on the crop and planting date. This is a significant advantage provided by minimally heated greenhouses.

In general, it's better to use a drip system to irrigate a range of vegetable crops in winter. It keeps the foliage from getting wet and reduces the likelihood of diseases, which develop when the air inside shelters is cold and damp.

When irrigating different crops, knowing when to water and for how long depends largely on the weather and the time of year. In general, especially in unheated shelters, no irrigation is needed in December and January, when day length is at its shortest. At this time, soil moisture retention is high, and the plants, which are in a phase of extremely slow growth, draw very little water up through their roots. When February rolls around and day length increases, it's time to come out of hibernation and make sure the soil doesn't get too dry. We monitor the crops and feel the soil several times a week. A moisture meter can also help determine if a bed is ready to be watered.

The best way to know if you need to irrigate is to measure soil hydration levels with a moisture meter. If you are considering irrigating a crop, the soil surrounding the plants should be more dry than wet.

In winter, in a minimally heated greenhouse, we irrigate once or twice a week for 25 minutes. We also try to water crops on sunny days, when plant transpiration and water requirements are higher.

Another crucial factor to consider when watering winter crops is the relationship between irrigation and a plant's frost resistance. Plants with a slight water deficit have been shown to better tolerate freezing temperatures because with less water in their cells, they are less likely to form ice crystals in below-freezing conditions. Since crystals are deadly, it's important to consider this when establishing an irrigation strategy before a major frost event. It's always safer to tend towards watering too little rather than too much, especially when nights drop below freezing in December and January.

Fertilizing Soil Organically in a Cold Climate

As is the case with irrigation, plant fertilization is different in winter. The ground is colder, which has a significant impact on the microbiological activity of the soil. This factor must be considered when making a fertilization plan because the accessibility of nutrients contained in organic fertilizers depends on soil biological activity.

Remember that microorganisms and fauna (earthworms, wood lice, fungi, bacteria, etc.) living in the soil break down organic matter found in our amendments, from compost and cover crops to ramial chipped wood and other organic fertilizers. They use the nutrients in organic matter to nourish their own metabolisms. In the process, they release more mineral nutrients into the soil solution—nitrogen, phosphorus, potassium, and other micronutrients—that become available for uptake by plants for their own nutrition. This symbiotic relationship between soil microorganisms and plants is the foundation of organic vegetable fertilization, but this process depends on one factor above all else: warm soil. As a result, there is no point in applying summer recommendations for fertilizing winter crops. Instead, opt for more frequent, smaller doses and focus on the timing, applying fertilizers under the right conditions.

Organic fertilizers are applied in a bed.

Microorganisms (bacteria, fungi, microfauna and macrofauna) feed on this organic matter.

Winter activity is slower

Mineral nutrients in the organic matter are released into the soil solution.

Limited availability

Mineral nutrients are absorbed by the plants.

Process slowed down by the cold and lack of sunlight, which inhibit metabolism

Heating the Soil

Since heat and humidity promote microorganism activity, we try to warm the soil for winter fertilization. Scientific research has demonstrated a direct correlation between increased soil temperature and increased bacterial activity.

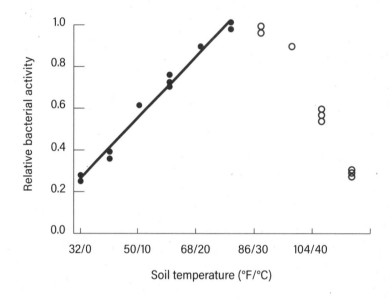

The relationship between temperature and bacterial activity
Source: Pietikäinen, Pettersson, and Bååth. "Comparison of temperature effects on soil respiration and bacterial and fungal growth rates."

As seen in this graph, bacterial activity increases continuously as the soil warms up, peaking at 77°F (25°C). Based on this principle, we know that increasing soil temperature by a few degrees gradually increases microbial activity and thus releases more nutrients into the soil. When they are warm, microorganisms are active: they feed on organic matter, which results in the mineralization of organic nitrogen. More nitrogen then becomes available for plant uptake, which stimulates growth.

Based on this information, our winter fertilization strategy follows three principles:

- Once a crop is transplanted or seeded, we do not add any fertilizer in November and December, when plant growth is slowest.
- If a crop is planted in September and harvested all winter, we fertilize it with compost and half the feather meal or pelleted poultry manure needed. The other half is added in February as light increases and growth resumes. For more precise dose information, refer to your local agricultural department recommendations.
- The exact doses, the quantity to be applied, and the frequency depend on several factors: the crop, weather conditions that promote vigorous plant growth, soil analyses, and residual nitrogen remaining from the previous crop.

Some studies (Amare and Desta, 2021; Gheshm and Brown, 2021; Snyder et al., 2015) show that you can warm the soil in a cold greenhouse by laying black mulch over the ground. On sunny days, this mulch will build up solar energy and transfer it to the soil through a conductive process. This heat transfer can increase soil temperatures by a few degrees (compared to a soil without any ground cover), which may explain why many farms in France use plastic mulch extensively in their unheated greenhouses.

At Ferme des Quatre-Temps, we have found that soil cultivation is the best approach to weed control for winter crops because it also aerates the soil and increases the mineralization of organic fertilizers. This is why we do not use ground covers in our unheated greenhouses.

An alternative to these ground covers: row covers placed over crops. They trap heat close to the ground, but we can still cultivate when needed.

Radiant heating systems can also be installed underground to warm the soil. A typical system consists of a network of pipes running underneath winter crops, with an antifreeze liquid (like glycol) circulating throughout. The liquid is heated, which generates the radiant energy needed to warm the soil, and growers can use a thermostat to control soil temperature. However, this method is still quite expensive. The efficiency and profitability of radiant heating in greenhouses are not yet proven.

CHOOSING FERTILIZERS AND DOSAGES

When the soil is warm and nutrients are more available, growers need to select fertilizers and dosages. Given the wide range of conditions on each farm, it would be risky for us to give precise guidelines for fertilizing winter crops. The best approach will depend on soil type and texture, previous crop fertilization, soil amendments and fertilizers, etc. Given previous explanations, we know that fertilization plans must focus on fertilizers containing nutrients that will quickly become accessible to plants.

When choosing fertilizers, there are many options. In our experience, growers should avoid overfertilizing winter crops with amendments that are high in nitrogen, phosphorus, and potassium. This can lead to high soil salinity and foster the development of soil diseases like pythium and fusarium, which are widespread in northern agriculture. It is especially true inside tunnels or greenhouses, where nutrient buildup is common because the soil is not exposed to the leaching effects of rainfall. Fertilizers that are well suited to winter crops are crab meal, alfalfa meal, feather meal, and compost.

Fertilization needs will vary between crops. As a rule of thumb, however, we split fertilizers into two or three applications, to better align with plant growth. The process of synchronizing fertilization and plant growth is particularly important in February, when crops start growing faster, as day length increases and the soil becomes warmer. By this time, fertilizers applied in the fall become relatively ineffective, so a small nutrient boost will be enough to reinvigorate plants nearing the end of a long winter. To check nutrient availability in the soil, a 2:1 test with a salinometer is a good reference. For some crops, like greens, applying a fertilizer after each harvest will help promote robust regrowth. Alfalfa or feather meal is an appropriate choice in this case.

Maintaining and Protecting Winter Crops

Many vegetables grown in winter will remain in the soil from September to March. This long growing season gives weeds the opportunity to become established, competing with crops for light, water, space, and nutrients. It's important to eliminate them as quickly as possible. To stay on top of weed control, our strategy at Ferme des Quatre-Temps is quite simple: we cultivate crops regularly.

While repeated cultivation is effective for weed control, it also aerates the soil, promoting plant growth. Year after year, we have seen this approach deliver positive results, and we recommend it to those who might want to follow in our footsteps. In chapter 6, we extensively covered the use of cultivation tools. As with summer production, the secret to effective weed control is using the right tool at the right time, with regularity and through quick interventions when weeds are still in the white-thread stage.

RODENTS

In winter greenhouses, vegetables aren't the only ones protected from the wind and cold. Small rodents also seek refuge in our shelters. Mice can quickly nibble vegetables, rendering a crop unsellable. To avoid losing an entire bed of radish or turnip, preventive action is the key to nipping this problem in the bud. Since poison is not an acceptable solution in organic farming, the best approach is to install traps combined with ultrasonic pest repellers plugged into the greenhouse.

After several experiments, we have found that the most effective trap can easily be made with a large plastic container and a few pieces of wood. We dig a hole in the soil, to the depth and diameter of the container, then fill the tub halfway with water. We coat the inner rim of the container with peanut butter and place small pieces of wood around the perimeter, creating bridges to the ground. The mice walk along these planks to reach the peanut butter and fall into the water. Make sure to check the traps almost daily to remove dead mice and readjust the set-up as needed. This simple technique is highly effective and inexpensive.

INSECT PESTS

After several years of running a four-season vegetable production in our greenhouses, we are facing new challenges in pest management. We have seen an increase in the populations of certain insects and are noticing earlier first sightings every spring, suggesting that growing vegetables in winter can increase insect pest pressure year-round. A preventive approach is therefore the key to effectively tackling these problems. Thrips and aphids are notable examples. Whatever the pest may be, crops must be regularly inspected to detect an emerging infestation as soon as possible.

At the first sign of aphids, we spray infested plants with a biopesticide and repeat the treatment until the problem subsides. It's essential to act quickly and follow the manufacturer's instructions as an aphid infestation can be extremely difficult to control once it becomes established. In the summer, we use beneficial insects to keep aphids in check, but these predators are too expensive to justify as a preventive winter measure. This is especially true because most beneficial insects are inactive or ineffective at cold temperatures.

Aphids are the main insect pests seen in winter greenhouses. It's important to know how to recognize them and to inspect crops at least once a week, to quickly detect aphids.

DISEASES

Plants are prone to fungal diseases throughout winter, mostly at the end of the season when crops have been in the shelter for several months. At this point, we start to notice that the vegetables are less healthy, which makes them more vulnerable to disease.

You must do everything you can to limit the presence of diseases in winter crops. While a disease may cause significant damage in its first season, it can go on to affect crops in subsequent seasons as well. This is why fungal spores cannot be allowed, under any circumstances, to build up in greenhouse soil.

Several simple preventive methods can help mitigate most risks. The work begins with crop planning; we choose disease-resistant cultivars and adjust plant spacing to promote better airflow. Next in our preventive approach, we prepare permanent beds in the fall, before planting winter crops. At this stage, we make sure to build up the raised bed again, with the rotary plow. By reforming the beds, we are able to improve soil drainage, which will reduce water retention and humidity around the crops.

Another best practice is installing perforated tubes to distribute air at ground level, around the plants. Whether the greenhouse is heated minimally or not, we leave our blower running continuously, pushing air through the tubes right onto the plants and preventing stagnant air pockets. We use these same tubes to distribute heated air around the plants during our dehumidification cycles.

Managing moisture in the greenhouse reduces the likelihood of fungal spore germination. When relative humidity is too high, we simultaneously heat (if possible) and ventilate the space several times throughout the day to reduce moisture levels. This dehumidification process is particularly important at dawn, when temperatures are dropping and relative humidity is at its peak. An unheated greenhouse can also be dehumidified with heat from the sun. Simply ventilate the greenhouse when solar energy begins to raise interior temperatures.

If, despite these efforts, a disease does develop, we remove any affected plants from the greenhouse to decrease risk of infecting others.

DISINFECTING GREENHOUSES

Greenhouse disinfection (i.e., a deep cleaning process carried out between seasons) is another method that, along with the previous strategies, can help tackle diseases and insect pests in winter. This work generally happens between winter and summer crops, in April.

Disinfection reduces the presence of pests and diseases in the greenhouse. The goal is to wipe the slate clean and start up the next crop with as few problems as possible.

Opt for disinfection when problems recur from season to season or from one year to the next. For example, if you are experiencing constant spider mite infestations, it is probably a good idea to disinfect your greenhouses. If, however, you are seeing a healthy balance between beneficial insects and pests, and between beneficial and harmful fungi, you may not need to disinfect. This distinction is important because disinfection wipes out many organisms and does not differentiate between good and bad. Friends and foes—all will fall. This is why we prefer not to interfere unless a greenhouse has recurring problems.

When disinfecting a shelter, the first step is to spray crops with a horticultural oil, such as PureSpray GREEN, which is certified for organic growing. Make sure to thoroughly cover the foliage. We also spray our greenhouse infrastructure and the ground with oil. Any pests here will suffocate and die. Once the plants are dry, remove them from the greenhouse. Next, make sure to take out all plant matter, whether crop residue or a weed. At this point, we also remove all production equipment: drip tape, pipes, hooks, string, etc. Because oil can damage greenhouse plastic, it's safer to rinse all plastic shortly after the oil treatment.

Now the deep cleaning begins. We hook up our pressure washer in the greenhouse and spray everything except the ground. The goal is to clean thoroughly, removing as much dust and debris as possible. After rinsing, we let it all dry before beginning to disinfect. On all greenhouse surfaces, we spray a disinfectant certified for use in organic agriculture. It's crucial to follow the instructions for the product you are using. A final rinse with water may be necessary.

The advantages of this disinfection process are that it is relatively simple and allows for a quick transition between growing seasons.

In all cases, prevention and good practices remain our best allies. Implementing the right methods from the start may be more time-consuming, but it will often save you headaches down the road—and you'll have the satisfaction of a job well done. Lastly, if problems with winter crops still recur, you may need to consult an agronomist, even during this quiet season.

Diseases in Winter Lettuce

In the winter, we monitor lettuce for downy mildew (*Bremia lactucae*), which develops particularly well when the weather is cool and damp. It is identified by yellow or brownish angular patches on the leaves. To protect against this disease, use drip irrigation to keep the lettuce dry. If there's no water on the leaves, mildew spores are less likely to germinate.

In greenhouses, winter lettuce can also be affected by powdery mildew (*Erysiphe cichoracearum*), which develops as a white mycelial growth on the surface of the leaves and compromises quality. Affected plants must promptly be removed from the greenhouse, and a phytosanitary treatment may be required to regain control.

Gray mold (*Botrytis cinerea*) is one of the diseases that cause the most damage to winter lettuce. The risks associated with it are significant: it spreads rapidly in low-light conditions and causes collar rot (gradually spreading to the entire head). It is identified by brown mold at the base of the plants, and a generalized and rapid dieback.

You can prevent gray mold by using precautionary growing methods: climate control (humidity control), transplanting lettuces early in the fall when growing conditions are good, heating the greenhouse in sunny weather, weeding, etc.

Fall Harvests
for Cold Storage

❄

What good is the warmth of summer, without the cold of winter to give it sweetness?

— John Steinbeck

Our ancestors knew how to store vegetables without modern refrigeration systems. We tend to think that they were limited to eating only carrots and potatoes in the winter. But the truth is they had access to many other vegetables, such as winter squash, cabbages, parsnips, rutabagas, beets, onions, leeks, and celeriac throughout the season. To enjoy the diversity and abundance of fall crops, our ancestors stored all these vegetables in root cellars. Today, the remains of these rudimentary refrigerators can be found scattered across the Quebec countryside.

In the 18th and 19th centuries, storage vegetables were kept in root cellars that farmers dug into the ground. This inexpensive way to protect harvests from winter frosts, thanks to the soil's insulating properties, was developed by Europeans who had traveled to North America. They were heavily influenced by the ancestral Indigenous practice of burying foods in the fall to protect them from frosts.

Root cellars are ingenious, simple, and effective solutions. Although this approach to food storage was gradually abandoned with the advent of modern refrigeration technology, the principle remains relevant. Planning meals around seasonal ingredients allows us to eat vegetables that have developed their full flavor. And storage crops are no exception. Because vegetables harvested in the fall are staples of a northern diet, it's worth paying special attention to them. It's not uncommon to hear a chef rave about the outstanding flavor of a storage crop harvested after the first frosts. Plants concentrate sugars in their cells when exposed to cold temperatures; this is a uniquely northern phenomenon and is one of the reasons that Ferme des Quatre-Temps products are outstanding.

Small farms should make a point of showcasing the remarkable taste of late-season crops. Visually appealing, storage crops are the perfect complement to freshly harvested vegetables grown in winter. In our experience, this complementary offering is integral to bringing winter crops to market and will even guarantee the success of a season.

That said, growing multiple storage crops is no simple endeavor. The main challenge is that most of these crops must be planted when the summer season is in full swing, when to-do lists are already miles long and field space is fully occupied. The only way to work around this is to create a plan ahead of time.

Planning a Fall Garden

At Ferme des Quatre-Temps, we start thinking about our fall crops in December. This is when we review the past year's highs and lows, to figure out what we will need in the coming year. Many questions come up in this reflection process:

- Which vegetables were most successful? Why?
- Which storage crops ran out during the winter?
- How can we regulate the supply?
- Were any vegetables unable to reach maturity before the first frosts?
- Do we want to experiment with new varieties?

This analysis allows us to establish a definitive list of vegetables that we want to produce, and the quantities needed. We then determine how long each crop will occupy a space in the garden. To calculate this, we take the number of days to maturity for each vegetable, then add them to the length of the harvest window (generally two or three weeks). One of our objectives in this process is to align the first harvest dates with the first autumn frost. At Ferme des Quatre-Temps, in Hemmingford, southern Quebec, the first frost usually hits around October 5.

▲ The start of winter is the best time to plan the next season, as well as the fall garden, while everything is still fresh in your memory. We review our notes from the past season and, most importantly, our sales reports for each vegetable.

To better understand the process, let's use winter radish as an example. This unique and tasty storage crop can handle a fall frost, without suffering any damage. In our experience, the ideal harvest date is around October 15. Since this crop requires 50 days to maturity (DTM), we sow it in trays on July 25. We then transplant it into the field 28 days later, on August 22, so that we can harvest it around October 15.

◄ Winter radishes are delightfully crunchy and add a colorful splash to our baskets and winter markets, which our clients love. Among our storage crops, they are one of the showstoppers.

To determine the seeding or field transplant date for each vegetable, take the first fall frost date and subtract the number of days to maturity for this crop. This gives you the planting date, i.e., when the crop must be direct seeded or transplanted into the field. Here is the formula:

**Date of first frost – (Days to maturity + Harvest window) =
Direct seeding or field transplant date**

For transplanted crops, you also need to consider the days spent in seeding trays before the transplant date.

We do this for every storage crop we intend to produce. This method gives us an overview, allowing us to plan which crops to grow, when to plant them, and which beds need to be freed up. To make things more concrete, picture three plots with ten permanent beds each, all set aside for the fall garden. We map out our fields, with each column representing a permanent bed and each row representing a two-week period. Every crop is then assigned to a box that is positioned both in time (horizontal axis) and in occupied space (vertical axis).

After assigning all fall crops a spot in the plan, we fill in the remaining spaces at the beginning of the season with quick-growing summer crops or cover crops. We expect a two-to-three-week gap between a crop's last day and the next planting date in the same bed. This allows us to turn in the first crop with a flail mower and cover it with a silage tarp. During this period, soil microorganisms break down the plant residue, leaving behind a bed surface that is ready to receive a fall crop planting.

Unlike biointensive market gardening, the traditional approach to vegetable growing does not necessarily seek to optimize production on every plot of arable land. To succeed, the biointensive model requires rigorous planning and a strategic attitude, so that growers can maximize revenue, while deftly managing crop successions throughout the summer season. Home gardens call for a similar approach. For more information, see Appendix 3, Starting a Fall Garden at Home (p. 244).

Example: Fall Garden Plan

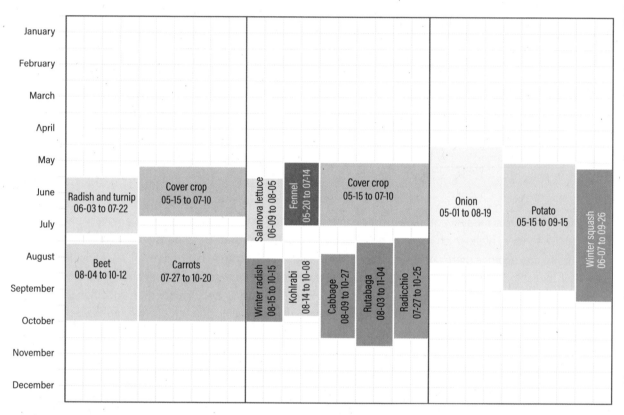

Radish and turnip 06-03 to 07-22	Cover crop 05-15 to 07-10
Beet 08-04 to 10-12	Carrots 07-27 to 10-20

Salanova lettuce 06-09 to 08-05

Fennel 05-20 to 07-14

Cover crop 05-15 to 07-10

Winter radish 08-15 to 10-15

Kohlrabi 08-14 to 10-08

Cabbage 08-09 to 10-27

Rutabaga 08-03 to 11-04

Radicchio 07-27 to 10-25

Onion 05-01 to 08-19

Potato 05-15 to 09-15

Winter squash 06-07 to 09-26

Storage crops grown in permanent beds.
Every plot is 35 feet (10 m) wide and 100 feet (30 m) long, and includes 10 permanent beds.
Each bed is 100 feet (30 m) long and 30 inches (75 cm) wide, with a 1-foot (30 cm) aisle.

Vegetables for a Fall Harvest

At Ferme des Quatre-Temps, we prioritize growing all root vegetables in the field, if they can be harvested in late fall. This is the case with carrots and beets. While they could grow in our shelters over the winter, we've found that they do quite well in the field, despite the cold autumn weather. Once harvested, they will keep for months in cold storage. We prefer to use our precious shelter space for vegetables that will be harvested over the winter and that do not have long-term storage potential.

Storage crops have specific characteristics, and the following information on each one provides details to help you better plan your fall harvest. It is worth mentioning that storage vegetables are generally harvested at a larger size than their summer counterparts, and they are usually stored dirty, which increases their potential shelf life.

The term "days to maturity" refers to the number of days that a vegetable needs to grow in the field, after a direct seeding or transplant, until ready to harvest. Starting with the days to maturity listed by the seed producer, you'll have to adjust them according to your field notes from past years.

Planting dates will vary, depending on how far north the farm is located. The dates shown here can help guide your choices, but they must be tailored to your local climate.

Beets

Although our way of producing beets is uncommon, it is optimized to meet the reality of an efficiency-oriented market gardener. We start them in our nursery before transplanting them into the field. With this method, we can avoid inconsistent germination rates and grow beets in uniformly filled beds. We also skip thinning, by transplanting a crop with perfect spacing.

Beets are harvested in the fall, after the first light frosts. To increase storage potential, we cut off the leaves in the field and store only the root, which we wash on demand.

Cultivars

Bull's Blood
  Chioggia
Cylindra
Rhonda
Touchstone Gold

Spacing

3 rows
5 inches (13 cm)

**Start in
indoor nursery**

35 days in a
128-cell tray

Transplant date

Early August

Days to maturity

50 days

**Storage
temperature**

32–34°F (0–1°C)

Cabbage

For our fall harvests, we select cultivars that are cold hardy and bred for storage. We prefer cultivars that grow a small head because they are a better fit for our clientele's needs.

Cabbage must absolutely be covered with insect netting as soon as it is transplanted, otherwise the crop might be lost to leaf-eating caterpillars, like cabbage looper larvae, diamondback moth larvae, and imported cabbage worms. This netting also helps prevent losses caused by gall midges (Cecidomyiidae), which are very difficult to control by any other means, especially in organic farming.

We harvest cabbage with a knife when the head has a suitable shape and density. At this point, we remove loose leaves around the head, making sure to keep two or three wrapper leaves to protect it while in storage.

Since cabbages are very cold hardy, they can be harvested after the first frosts.

Cultivars

Caraflex
Ruby Perfection
    Storage No. 4
Tiara

Spacing

2 rows
18 inches (45 cm)

**Start in
indoor nursery**

28 days in a
72-cell tray

Transplant date

Early August

Days to maturity

75 days

**Storage
temperature**

32–34°F (0–1°C)

Carrots

Our storage carrots harvested after the first frost are so sweet that they are a hot topic of conversation at the market. They are so popular that we always run out, and we try to grow as many as we can.

We seed them using a Six-Row Seeder, and do not thin the crop. Carrots need to be seeded in relatively clean soil (i.e., without weeds) because they will inevitably require a few manual weeding sessions. Our tip: we sow fall carrots in our garlic beds after harvesting the crop. Because these beds are covered with straw for several months, the following crop will experience less weed pressure.

To reduce our workload, we harvest storage carrots over a few weeks, starting in November. They can withstand a hard frost and can stay in the field until it starts to snow. We use a broadfork to loosen the soil, which makes harvesting easier. We then cut off the tops in the field and store the carrots unwashed in a cold room, cleaning them as needed for sales.

Cultivars

Boléro
Mokum
Napoli

Spacing

6 rows
2 inches (5 cm)

Seeding date

End of July

Days to maturity

70 days

Storage temperature

32–34°F (0–1°C)

Celeriac (Celery Root)

Celeriac is becoming more popular every year. For people who love its unique flavor, it is a staple among storage vegetables. Whether roasted with carrots, sautéed in a soup base, or served raw in a remoulade, this vegetable will fit a range of hearty dishes!

We transplant celeriac into geotextile fabric because it stays in the garden for almost the entire season. It requires very little maintenance, aside from regular watering and a foliar spray containing boron and calcium, to prevent a deficiency resulting in heart rot.

Once satisfied with the size of your celeriac plants, you can harvest them at any time in the fall. Make sure to do this before a significant frost event. When harvesting celeriac, we remove all roots and leaves, brushing off as much soil as possible. We don't wash it, however, so that it will keep better in cold storage.

Cultivars

Brilliant
Mars

Spacing

3 rows
12 inches (30 cm)

Start in indoor nursery

70 days in a
128-cell tray

Transplant date

Early June

Days to maturity

95 days

Storage temperature

32–34°F (0–1°C)

Chinese cabbage

Chinese cabbage is the first Asian vegetable in the brassica family to have been cultivated in North America. It grows quite well in cool fall temperatures. Among our customers—who love this vegetable—it no longer needs any introduction. If stored under the right conditions, it will stay fresh for 3 months. We recommend keeping Chinese cabbage in closed bins, in a high-humidity room at 32°F to 39°F (0-4°C).

We grow Chinese cabbage using the same methods as those described for standard cabbage. It must also be covered with insect netting to prevent damage caused by various caterpillars. After this crop is transplanted at the end of August, it will require about 60 days to reach maturity.

Chinese cabbage is ready to be harvested when the head feels dense. Give it a squeeze, and if you feel little resistance, the cabbage is not dense enough and will need a few more days in the ground.

Cultivars

Bilko
Rubicon

Spacing

2 rows
18 inches (45 cm)

**Start in
indoor nursery**

28 days in a
72-cell tray

Transplant date

Early August

Days to maturity

60 days

**Storage
temperature**

32–34°F (0–1°C)

Garlic

Garlic will keep quite well if harvested at the right time and cured properly. It is an excellent complement to our winter vegetables. Overall, demand for local garlic is strong because its taste and quality are head and shoulders above imported garlic.

This crop requires additional preparation because garlic is sown in the fall then harvested in the summer. After planting garlic, we cover it with a layer of straw, which provides insulation and protects it from below-freezing winter temperatures. Over the following spring and summer, it will grow through the straw. We harvest it once the plant has grown a scape and two to three leaves have begun to dry. We also harvest the scapes (garlic flowers), which we sell fresh in bunches of ten.

To cure garlic, we cut off the stems, then let it rest for 2 to 3 weeks in a well-ventilated space, until the outer skins are completely dry. We then store it in open bins kept in a low-humidity room, at 15°C, for the entire winter.

Cultivar	Days to maturity
Music	300 days

Spacing	Storage temperature
3 rows 5.5 inches (14 cm)	50–59°F (10–15°C)

Seeding date

Mid-October

Kohlrabi

Although we grow kohlrabi all summer, it tastes best in the fall. We grow Kossak and Superschmelz cultivars because they have good texture, fantastic flavor, and, above all, excellent cold storage properties.

We transplant kohlrabi in August and cover it with insect netting to protect young plants from flea beetles and caterpillars.

We harvest kohlrabi about 2 months later, when plants are 8 inches (20 cm) wide, before or during the first light frosts in the fall. At this diameter, kohlrabi is not fibrous, will stay tender, and can be eaten raw or cooked.

When we harvest kohlrabi and move it to a storage space, we make sure to handle the produce with care. Impacts can create cracks in the bulbs, which will decrease their storage potential and render them unfit for market.

Cultivars

Kossak
Superschmelz

Spacing

3 rows
12 inches (30 cm)

**Start in
indoor nursery**

28 days in a
72-cell tray

Transplant date

Early August

Days to maturity

60 days

**Storage
temperature**

32–34°F (0–1°C)

Radicchio

Radicchio is a brightly colored red or green chicory variety that forms a tight head. Several cultivars are available and all are quite different from one another. Radicchio is the perfect leafy vegetable for a fall harvest because it thrives in cold weather and will keep well in a cold room. Some cultivars will even keep until Christmas. We transplant it into the field at the end of summer and harvest it during winter frosts, in late fall.

Restaurateurs love the bitter taste of radicchio, and our customers are increasingly familiar with this vegetable. As a specialty product, it makes our winter offering stand out.

Cultivars

Fiero
Leonardo
Virtus

Spacing

3 rows
12 inches (30 cm)

**Start in
indoor nursery**

25 days in a
128-cell tray

Transplant date

Late July to early
August

Days to maturity

65 to 70 days

**Storage
temperature**

32–34°F (0–1°C)

Rutabagas

Rutabagas—also known as swede, Swedish turnip, and sometimes simply turnip—are among the vegetables that past generations have traditionally stored over winter. We transplant them into the field at the end of summer, 2 to 3 weeks before winter radishes, and harvest them after the first frosts. As with winter radishes, we remove the leaves, to maintain quality throughout the cold storage season.

Rutabagas can be sold throughout the winter. With the right storage conditions, it will still be good to eat come spring.

Cultivars

Joan
Gilfeather
Laurentian

Spacing

3 rows
8 inches (20 cm)

**Start in
indoor nursery**

28 days in a
128-cell tray

Transplant date

Early to
mid-August

Days to maturity

80 days

**Storage
temperature**

32–34°F (0–1°C)

Storage onions

With storage onions, bulb development depends on day length (photoperiod). Varieties that are typically available north of the 35th parallel require a day length greater than 14 hours to develop a bulb. This is why onions need to be transplanted into the field in early spring (late April or early May).

Unlike other crops grown in closely spaced rows, onions don't produce enough foliage to compete with weeds. Seedlings should not be placed where weeds went to seed in previous years. If a bed is not clean, the best option is to set up a woven ground cover before transplanting.

Storage onions come to maturity 10 to 12 weeks after planting. At this stage, the top folds over onto itself, which means the plants are ready to begin the drying process. Once 70 percent of the foliage has fallen to the ground, we manually break the onions that are still standing so that their necks will begin to close.

We dry them for 2 to 3 weeks, with the stems still attached, on wire tables located in a well-ventilated space.

At the end of this process, we check whether the neck of each onion is tight (has closed properly), a sign that it has finished curing and will keep well in storage. At this point, we can cut the stems and store the bulbs all winter in a dry room kept at 60°F (15°C).

Cultivars

Cortland
Red Carpet
Redwing
Rossa di Milano
Talon

Spacing

3 rows
10 inches (25 cm)

**Start in
indoor nursery**

45 days in an
open seed tray

Transplant date

Early May

Days to maturity

80 days

**Storage
temperature**

32–34°F (0–1°C

Storage potatoes

Storage potatoes are a staple of the winter diet. We sow them in the spring for a harvest at the end of summer. Two weeks before the field planting date, we expose seed potatoes to light and to roughly 50°F (10°C) temperatures to induce presprouting. This technique gives us a head start and delivers good germination rates.

Throughout the season, make sure to carefully monitor the crop for Colorado potato beetles (CPBs). This insect feeds on the potato plant foliage, which can lead to crop losses. At the first CPB sighting, manage the pests by shaking or brushing insects into a bucket of slightly soapy water. It's important to prevent infestations while the plants are in bloom because this is when tubers are also developing and growing. In this stage, foliage is essential.

Plan to hill potato plants twice during the season: once when they are 6 inches (15 cm) tall and again at 12 inches (30 cm).

Potatoes are ready for harvest when the foliage is wilting or yellowing. With the help of a fork, we gently pull them out of the ground.

Cultivars

Chieftain
French Fingerling
German Butterball
Prairie Blush
Russian Banana
Yukon Gold

Spaing

1 row
8 inches (20 cm)

Seeding date

Mid-May

Days to maturity

110 to 120 days

Storage temperature

40–45°F (4–7°C)

Winter radishes

Winter radishes come in a stunning array of colors. From the bright pink Red Meat to the purple KN-Bravo and the two-toned Green Luobo, these cultivars add a splash of color to the winter landscape!

We prefer to start them in our nursery and transplant seedlings into the field at the end of the summer, to limit time spent weeding and to ensure uniform plant growth throughout the beds. In soils with a history of boron deficiency, a boron spray may be needed to prevent the development of brown hearts and to maintain root quality.

We harvest radishes in late fall, and cut off the leaves right at the base, leaving them in the field and storing only the roots.

Cultivars

Alpine
Green Luobo
KN-Bravo
Nero Tondo
Red Meat
Red King

Spacing

3 rows
8 inches (20 cm)

Start in indoor nursery

28 days in a
128-cell tray

Transplant date

Early to
mid-August

Days to maturity

50 days

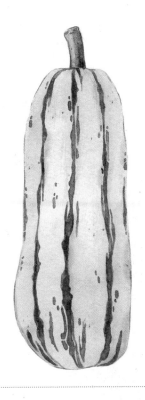

Winter squash

This fall crop provides the most variety for our customers. Butternut, Delicata, Honeynut, Acorn, Red Kuri, and Spaghetti are cultivars that we grow side by side.

In early summer, once the frost danger has passed, they are transplanted and will spend the entire season in the field. For this reason, we grow them with geotextile fabric covering beds and aisles, with drip irrigation installed underneath. This setup keeps weeding requirements to a minimum and is the key to a lucrative winter squash crop.

After it is transplanted, winter squash must immediately be covered with insect netting to keep out striped cucumber beetles and squash bugs, which can wreak havoc in a crop. We remove the netting once the flowers have bloomed, to allow for pollination.

We harvest winter squash in late fall, but this must happen before any major frost event. Temperatures below 28°F (–2°C) can damage them and significantly shorten their shelf life.

Next, most squash will need to cure for 2 to 3 weeks. At this stage, we store them in a heated room kept between 80°F and 90°F (27–32°C), with constant ventilation. This process helps toughen up the skin and promotes healing of any wounds. Once the squash have cured, we keep them in a dry room, at 60°F (15°C).

Cultivars

Acorn
Butternut
Delicata
Honeynut
Red Kuri
Spaghetti

Spacing

1 row
36 inches (90 cm)

Start in indoor nursery

14 days in a 72-cell tray

Transplant date

Early to mid-June

Days to maturity

85 to 100 days

Storage temperature

50–59°F (10–15°C)

207

Harvesting Storage Vegetables

Harvesting storage vegetables requires an intuition that is shaped by years of experience. Every season is different, and this will be even more true in the future, as a result of climate change. The secret is to thoroughly assess temperatures and the weather to decide when to harvest each crop, based on its maturity and the potential consequences of a delayed harvest. If you postpone, the weather will get colder, and frost will be more likely to damage crops. On the other hand, if you harvest too soon and the vegetable is not fully mature, it may not keep as well and will have a less developed flavor.

For successful fall harvests, the first step is to determine each vegetable's cold hardiness, and then carefully monitor the weather for upcoming frost danger. You'll need to be ready to harvest a crop at the drop of a hat, to avoid incurring losses. Some vegetables can also be shielded from the cold with row covers.

Once they are ready, vegetables are harvested without their foliage to increase their storage potential. We cut the tops off in the field, returning precious nutrients to the soil (nitrogen, phosphorus, potassium, etc.).

Any foliage left on vegetables after harvest will draw water from the root, which will affect the texture and weight of the vegetable. There is also a chance that the foliage will rot and trigger a chain reaction, causing the entire vegetable to spoil. This is why it's also important not to harvest produce damaged by handling or insects. Throughout this process, handle vegetables with care because any impact has the potential to cause damage. The result: these vegetables will quickly deteriorate in storage and have a shorter shelf life.

Before delivering root vegetables, like carrots, beets, and winter radish, we run a weekly washing session, using a root vegetable washer, which is barrel-shaped. We simply put the vegetables in the drum and turn on the washer. As it rotates, water circulates throughout, to quickly and efficiently clean root vegetables. If you want to sell vegetables year-round, this step will require access to a heated washing station (for staff comfort and access to water).

Best Practices for Storing Fall Crops

There is an art to storing vegetables over extended periods of time, given the many interacting factors: temperature, humidity, gases, storage diseases, damage from the harvesting process, etc.

To maximize our odds of success, we follow the best practices recommended for each crop. This allows us to limit storage losses, though we cannot eliminate them altogether. We keep most of our storage vegetables in a root cellar or in a cold room. They can be stored in other spaces, as long as they provide optimal conditions.

TEMPERATURE

When storing crops in winter, the most important factor is frost protection. Vegetables exposed to temperatures that are too cold will develop internal ice crystals that tear through their cells. Once thawed, they take on a wet appearance and rot quickly. To avoid losses, keep produce in well-insulated spaces, such as cold rooms, root cellars, garages, and basements.

Once you've selected an insulated room, make sure to integrate a cooling and heating system. It may seem counterintuitive to heat a cold room, but if temperatures are –4°F (–20°C) outside, it might get too cold inside. The storage space should never drop below 32°F (0 C) as this could cause irreparable damage to the stored crops. A heating system is therefore necessary, especially if the room is poorly insulated.

The cooling system can be as rudimentary as a CoolBot temperature controller. This inexpensive solution allows users to cool a cold room with a simple household air-conditioning unit. For bigger facilities, consider installing a professional-grade refrigeration system. If your storage space is already kept at the required temperature, you won't need an additional temperature control system.

The optimal temperature depends on which vegetables you are storing since each has an ideal storage temperature. Given the storage facilities typically seen on diversified vegetable farms or in homes, we have split storage vegetables into two categories to streamline operations: cool temperatures (50–59°F [10–15°C]) or cold temperatures (32–39°F [0–4°C]).

Winter squash and garlic keep best in cool temperatures, while root vegetables (carrots, beets, winter radish, etc.), cabbage, onion, potato, celeriac, radicchio, and kohlrabi favor cold temperatures.

HUMIDITY

Moisture management in a storage space is essential to ensure the profitability of a winter season and to maintain the quality of your produce. Poor humidity control can quickly lead to significant losses; for instance, vegetables become dehydrated when relative humidity is too low. For products sold by weight, this drying process will result in a loss of revenue. Let's say that vegetables become 20 percent dehydrated: if you were expecting $5,000 in weekly sales, this would represent a $1,000 loss! What's more, excessive drying affects the texture of stored vegetables, making them less appealing to customers. Although a decrease in mass is inevitable since vegetables lose water through transpiration, drying can be slowed by using closed containers like bins or plastic bags. A range of techniques may also be used occasionally to increase relative humidity in a storage space, including spraying water onto the floors.

However, if relative humidity is too high, it will increase the likelihood of rot and fungal diseases. Storage vegetables should be carefully monitored, and in the event of a disease outbreak, humidity levels can be decreased.

GAS EXCHANGE: ETHYLENE AND CO_2

Ethylene is a gas naturally released in large quantities by some vegetables and fruits, such as tomatoes, that are referred to as "climacteric." This growth-regulating gas accelerates ripening through increased respiration. In a storage space, it can cause significant damage (rot, bitter flavor, yellowish leaves, softened and shriveled produce, etc.) when it comes in contact with non-climacteric fruits and vegetables.

To limit ethylene damage, make sure to isolate climacteric products and avoid storing such fruits as apples, pears, and melons with your vegetables. Good air circulation will also reduce ethylene and CO_2 buildup in a storage space.

CONTAINERS AND INVENTORY MANAGEMENT

To store fall vegetables, a range of containers will do the trick: perforated or solid plastic bags, mesh plastic or fabric bags, wooden crates, plastic totes with a lid, etc. The best container will depend on each vegetable's moisture requirements as well as the quantity to be stored. For example, cabbage keeps well when stacked in a wooden crate, while radicchios will keep best in plastic bags or closed plastic bins.

Once all fall crops are in the storage space, inventory management begins. Temperature and humidity must be monitored and adjusted constantly. If vegetables freeze slightly, they may be saved by gradually increasing the temperature to above freezing. Plan to sort crops every two weeks. Quickly remove any damaged vegetables before they affect neighboring produce. Closed bags or bins limit potential losses by isolating small batches of vegetables.

A Guide to Storing Fall Crops

This table is a helpful guide to preparing storage spaces, so you can store vegetables throughout the winter. It will need to be tweaked according to the crops you grow as well as the temperature and relative humidity (RH) that you maintain in your storage space.

Crop	Temperature	Container	Duration	Notes
Cold and humid (95-99% RH)				
Beet	32–34°F (0–1°C)	Perforated plastic bags or plastic bins with a lid	3-5 months	Remove foliage before storing
Carrot	32–34°F (0–1°C)	Perforated plastic bags or plastic bins with a lid	6 months	Keep an eye out for storage diseases like *botrytis* (gray mold) and *scierotinia* rot (white mold)
Celeriac	32–34°F (0–1°C)	Perforated plastic bags or plastic bins with a lid	6 months	Trim roots again before market if damaged in storage
Cabbage	32–34°F (0–1°C)	Wooden crates or bulk piles	6 months	Leave 2 or 3 wrapper leaves around the head; monitor for any sign of storage diseases like *botrytis* (gray mold) and *scierotinia* rot (white mold)
Chinese cabbage	32–34°F (0–1°C)	Perforated plastic bags or plastic bins with a lid	1-2 months	Remove any damaged leaves before market

Crop	Temperature	Container	Duration	Notes
Kohlrabi	32–34°F (0–1°C)	Perforated plastic bags or plastic bins with a lid	4 months	Cut off the leaves near the base of the bulb
Storage potato	39–45°F (4–7°C)	Wooden crates, bulk piles, or paper bags	4–5 months	Do not expose tubers to light
Radicchio	0–2°C	Perforated plastic bags or plastic bins with a lid	4–5 weeks (Fiero and Leonardo) 3–4 months (Virtus)	Before market, remove any leaves damaged in storage
Winter radish	32–34°F (0–1°C)	Wooden crates, bulk piles, or paper bags	2–4 months	Remove foliage before storing
Rutabaga	32–34°F (0–1°C)	Wooden crates, bulk piles, or perforated plastic bags	4–6 months	Remove foliage before storing
Cold and Dry (65–70% RH)				
Storage onion	37–41°F (0–1°C)	Mesh bags or cardboard boxes	6 months	Sort regularly to remove damaged and rotten bulbs
Cool and Dry (50–65% RH)				
Garlic	50–59°F (10–15°C)	Mesh bags or cardboard boxes	3–6 months	Remember that a crop's storage potential and quality depends on the success of the drying process
Winter squash	50–59°F 10–15°C	Bulk piles or wooden crates	1–5 months (depending on the cultivar)	Sort regularly to remove rotten squash if stored in wooden crates

Growing and
Selling Vegetables
Year-Round

❋

Strong customer relationships drive sales, sustainability, and growth.

— Tom Cates

By the time fall harvests are over, when we've packed away all our storage crops and the field is blanketed in snow, the winter growing season is already well underway. Because we want to offer an appealing array of produce to our customers, we shift our energy to focus on greenhouses and tunnels where we grow about thirty kinds of fresh vegetables. With this combination of storage crops and winter greens, we can generate solid sales and keep customers highly interested in local food. It also provides an opportunity for us to bridge the gap between the last summer vegetables (sold in the fall) and their return in the following spring. This approach allows us to bring produce to market year-round, which is a powerful tool in building customer loyalty.

Marketing and Selling Vegetables in Winter

Bringing our vegetables to market is particularly exciting in winter. With a smaller production team, we make sure to sell only the highest-quality products. Restaurateurs and other customers are amazed to see so much local diversity. Far from austere, our winter vegetable offering presents a vibrant color palette, chock-full of enticing greens and other vegetables!

We primarily sell our winter products to the public directly, through an online store that we run on the farm. For people looking to order a basket, the process is simple. They log into the system and select the vegetables from a list of products. We make sure to bring variety to our weekly selection, updating our lists according to the latest crops that are available in our shelters.

This approach to marketing and sales is simple and requires little effort. We only harvest the quantities that have been sold and deliver baskets once a week to a few drop-off locations. Since we don't know the number of weekly orders in advance, we have less information for crop planning. Fortunately, the demand for local vegetables in winter is so high that we don't often end up with a surplus and are more likely to run out.

An integral part of our marketing and sales strategy is reliably bringing products to restaurateurs with whom we maintain a strong year-round business relationship. They are increasingly seeking locally sourced food and will opt to buy vegetables grown in their area, if available. After several years of working with this customer base, we have learned some lessons.

First, a consistent vegetable supply is essential. When offering a new product, we try to make sure it will be available for several weeks. Many restaurants build their menu based on the vegetables we supply and rely on continuous deliveries so they can offer a new dish over multiple weeks.

Second, outstanding quality is key. We carefully select each vegetable to package standardized (size and quantity) products across all deliveries. By paying attention to these details, we can significantly streamline the work happening in kitchens, because cooks and chefs will know exactly what to expect.

Every week, we post our list of available produce on Monday morning. Restaurants must submit their order by Tuesday, using our online platform, then we deliver the vegetables on Wednesday. We never deviate from this routine, and clients appreciate our reliability. On delivery days, we set aside time to talk to our customers and get feedback about our products. Were the radishes too big? Are some vegetables available only in quantities that are too small? We pass this information along to the team and adjust our services accordingly.

While developing these business relationships is time-consuming, the reward is invaluable: seeing our produce featured in a restaurant's meals. In fact, there is a natural bond between growers and chefs: we share a deep love of beautiful vegetables.

◄ Duck magret, pan-fried gnocchi, seasonal vegetables, split soy sauce dressing, oat crumble

▲ Red and yellow beets, Joséphine goat cheese, braised endive with sea buckthorn

Chefs: Étienne Huot, Denis Vukmirovic, and Lyssa Barrera.
Photographs by Élizabeth Delage.

221

Once we've received the orders, they obviously need to be prepared with care. First, we use a root crop washer (barrel washer) to clean vegetables that have been stored dirty. We then carefully harvest greens in our shelters and wash them with the bubbler. This tub pushes air into the water, mixing the leaves and rinsing them at the same time. It is the perfect tool for assembling our mesclun. Once the greens are clean, we bag them and place them next to the storage vegetables, for current orders. Everything is delivered fresh, which our customers love.

Growing Year-Round

By the time we have finished harvesting the winter crops in our shelters and storage vegetables are running out, we are already well into the summer crop planting schedule. At this point, we have even begun to harvest summer vegetables. Seasonal overlap is necessary as it allows us to provide a continuous supply. This means that two things happen as soon as we finish the last harvest of a crop: we pull it out of the bed to make space for a new planting, and we begin harvesting that same crop from a previously planted bed that is now ready for market. These transitions are the result of a fine-tuned crop plan that we have perfected over the years.

JANUARY	FEBRUARY	MARCH	APRIL	MAY	JUNE

Early spring harvests

Winter crops

JANUARY	FEBRUARY	MARCH	APRIL	MAY	JUNE
• Tend to winter crops	• Harvest winter crops	• Transplant summer crops into greenhouses (tomato, eggplant, etc.)	• End of winter baskets	• First early spring harvest	• Transplant tomato, eggplant and field peppers into caterpillar tunnels
• Seed tomato, eggplant, and ginger into trays	• Misc. projects (tool upkeep, repairs, orders, etc.)	• Seed early spring crops in the nursery	• Start planting in tunnels or under row covers	• Summer farmers' markets begin	• Weed, tend to crops, harvest
	• Graft tomato and eggplant		• Open the field irrigation system	• Summer baskets begin	• Start scouting for pests in the field
			• Start the summer season with a bigger team	• First tomato harvest	• Sow the first cover crops
			• Watch for below-freezing temperatures		

With this workflow, we are able to grow crops over twelve months rather than six. As we try to meet the farm's financial goals, we end up being less dependent on summer crops because the work is spread out over a longer period of time. As a result, we can offer our team year-round jobs and get to truly experience the changing seasons.

Here's what our work cycle looks like when we break it down by month:

JULY	AUGUST	SEPTEMBER	OCTOBER	NOVEMBER	DECEMBER

Fall crops (September–November)

Summer crops (July–November)

Winter crops (October–December)

JULY	AUGUST	SEPTEMBER	OCTOBER	NOVEMBER	DECEMBER
• Sow winter carrots • Ongoing crop successions in the field • Garlic harvest • Keep up the harvesting pace	• Plant fall crops • Fall harvests • Transplant the last succession of greenhouse cucumbers • In the nursery, seed the first crops for unheated greenhouses	• Begin planting winter crops • Start season extension in the field (set up caterpillar tunnels) • Monitor weather for freezing temperatures	• Summer farmers' markets end • Summer baskets end • Fill the cellar with storage vegetables • Store field equipment and cover beds with silage tarps • Winter baskets begin • Plant garlic	• Crop planning for the next season • Order seeds and other supplies • Inventory and storage • Close down the nursery until January • Harvest and deliver winter baskets	• 3 weeks of vacation • Deep clean the packing room • Prepare grafting equipment

Conclusion

You never change things by fighting the existing reality. To change something, build a new model that makes the existing one useless.

— Buckminster Fuller

Nothing is more resilient than nature. By using it as inspiration and a basis for our methods, we can tap into the very source of this strength. Through careful study of natural mechanisms and years of experimentation, we now have the expertise to successfully grow vegetables year-round with a low-tech-high-return approach. It all starts with a better understanding of plants that are already well adapted to the cold: what limits their growth and what doesn't, and what protection they need to grow successfully even in harsh winters like those seen in northern climates. In doing so, we are choosing to work with nature, rather than against it! Everything then becomes possible.

This statement is nothing new, and vegetable growers have been working for centuries to produce crops in the winter, just as our ancestors knew how to preserve vegetables to enjoy a varied diet throughout the seasons.

But modernity, which is not always synonymous with progress or improvement, has tried to circumvent seasonal realities. Through means that include transportation, heating, and artificial lighting that recreate certain climates, industrial agriculture is dedicated to offering customers fresh tomatoes, red peppers, and other off-season vegetables year-round. Because we are creatures of habit, grocery stores carry these products in winter, even when the quality is vastly inferior, with a bland taste and uninspiring texture. This way of growing and consuming food, disconnected from the seasons, results from a system of industrial agriculture divorced from nature.

But for those of us who appreciate the bounty of our four contrasting seasons, it seems preposterous that anyone would want to artificially create a summer climate in the winter. Why would we even care to eat strawberries in January grown in high-tech vertical greenhouses without soil? Why even bother with such technology when the reality is that we have everything we need to feed our communities with high-quality, nutrient-dense vegetables year-round *without* such gimmicks.

When grown under conditions that respect the northern reality, vegetables develop incomparable flavor and quality. Ask any chef that collaborates with us at Ferme des Quatre-Temps, and they'll tell you straight up: our winter veggies are absolutely delicious, incomparable to anything else. Like the extra income from our winter production, such amazing feedback is truly rewarding for our hard work. So, in the end, it's time to get back to basics.

We believe there is nothing quite like being in sync with the seasons. Society is increasingly moving towards greater food autonomy, and we are witnessing a rising demand for access to local fresh products. We believe this emerging trend is an opportunity for us to reflect collectively. Before investing public funds into ready-made solutions, why not take a moment to ponder the best solution for growing and eating vegetables year-round in a northern climate?

Because we have been thinking about this question for years, the answer seems obvious. We believe our actions should contribute to changing our agricultural model. It is possible for our food system to adopt a model that relies on a local vegetable supply, but this requires time, energy, and resources to develop winter agriculture that bridges the gap between fall and spring.

In Quebec, we are witnessing a true agricultural revolution: in 2020, for the first time in fifty years, the number of start-up farms exceeded the number that had ceased operating in that same year. This is proof that it is possible to reverse the decline of our agricultural community. We have noticed that most of these new farms are rather small operations whose goal is to feed their communities instead of participating in the globalized agricultural market.

Though it has long been considered a marginal form of agriculture, this model selected by these family farmers has, more than ever, demonstrated the viability of a farm operating on a human scale. More and more of us now believe that the vegetable growers running these small farms are the future of agriculture, and we are speaking up. We hope that our experience with winter farming will serve as a guide, so that they can extend their offering over a longer period, allowing families to sign up for baskets on an annual rather than seasonal basis. In that regard, the Quebec revolution can serve as a model of what's possible: we can transform the agricultural landscape with a collective vision built around small, cohesive four-season farms that unite communities.

We also hope that our research and experience might help growers turn their vision into a reality, or at least inspire them. In any case, we wish them courage and success in their agrarian adventures.

To all our cherished colleagues, know that you are the true stewards of our food sovereignty. Many leaders and the people you feed recognize this and support you in a way that truly reflects your value.

As for us, we simply love to perfect our art and feel proud and encouraged to participate in improving this collective know-how.

— Catherine and Jean-Martin, Hemmingford, December 2022

Appendices

Appendix 1: Winter Crop Spacing

The following table shows our preferred spacings for winter crops grown in tunnels or greenhouses. Many differ from our standard summer spacings. The goal is to improve airflow and access to sunlight within the crops. When direct seeding, we use the Jang Precision Seeder. The notes and codes listed in the table therefore indicate settings to be used with this tool.

We also start many vegetables in our nursery before transplanting them into a greenhouse. This method is ideal for winter vegetables because one of our greatest challenges is transitioning between seasons inside our shelters. Since summer crops are often more profitable than winter crops, we want to keep them in the ground for as long as possible. By starting winter crops in the nursery, we can allow summer crops to remain in the greenhouse for an additional 3 to 4 weeks before removing them to make way for winter seedlings. This is a successful strategy for ensuring profitability, and it allows us to establish an optimal density in our transplanted winter crops.

The spacing recommendations in the table align with the latest conclusions drawn from our experiments. This is a good starting point for a winter crop plan. The data can be tweaked to meet the needs of each farm (soil type, planting dates, climate, access to light, etc.).

Crop	Number of rows	In-row spacing	Notes
Arugula	6	0.5 in. (1.3 cm)	• X-24, 14F-9R, ≠ • Brush: halfway
Asian greens (tatsoi and mizuna)	6	0.5 in. (1.3 cm)	• F-24, 14F-9R, ≠ • Brush: halfway
Baby kale	6	0.5 in. (1.3 cm)	• F-24, 14F-9R, ≠ • Brush: halfway
Baby Swiss chard	6	0.5 in. (1.3 cm)	• G-12. 14F-9R, ≠ • Brush: halfway
Bok choy	3	9 in. (23 cm)	• Sow in a 50-cell tray, 42 days before transplanting; when transplanting, intercrop with kale

Crop	Number of rows	In-row spacing	Notes
Carrot	4	1.5 in. (3.8 cm)	• X-24, 11F-13R, ≠ • Brush: halfway
Celery	3	12 in. (30 cm)	• Sow in a 50-cell tray, 70 days before transplanting
Chinese cabbage	4	6 in. (15 cm)	• Sow in a 50-cell tray, 42 days before transplanting
Cilantro	3	9 in. (23 cm)	• Sow in a 72-cell tray at 3 seeds per cell, 42 days before transplanting
Claytonia	4	1.5 in. (3.8 cm)	• YYJ-24, 13F-11R, ≠ • Brush: halfway
Dandelion	4	6 in. (15 cm)	• Sow in a 72-cell tray, 30 days before transplanting
Green onion	3	9 in. (23 cm)	• Sow in a 72-cell tray at 5 seeds per cell, 55 days before transplanting
Kale (mature)	2	12 in. (30 cm)	• Sow in a 50-cell tray, 42 days before transplanting
Komatsuna	4	6 in. (15 cm)	• Sow in a 72-cell tray, 30 days before transplanting
Lettuce	5	6 in. (15 cm)	• Sow in a 72-cell tray, 35 days before transplanting
Mâche	4	1.5 in. (3.8 cm)	• YYJ-24, 13F-11R, ≠ • Brush: halfway
Mustard	6	0.5 in. (1.3 cm)	• X-24, 14F-9R, ≠ • Brush: halfway
Parsley	3	12 in. (30 cm)	• Sow in a 50-cell tray, 49 days before transplanting

Crop	Number of rows	In-row spacing	Notes
Potato	2	12 in. (30 cm)	• Begin presprouting 1 month before planting; sow in staggered rows
Radish	6	1.5 in. (3.8 cm)	• F-24, 14F-9R, ≠ • Brush: halfway
Senposai	3	12 in. (30 cm)	• Sow in a 72-cell tray, 30 days before transplanting
Sorrel	5	4 in. (10 cm)	• Sow in a 72-cell tray at 2 to 3 seeds per cell, 30 days before transplanting
Spinach	4	6 in. (15 cm)	• Sow in a 128-cell tray, 28 days before transplanting
Swiss chard (mature)	2	12 in. (30 cm)	• Sow in a 50-cell tray, 42 days before transplanting
Turnip	6	1.5 in. (3.8 cm)	• Y-24, 11F-11R, = • Brush: halfway
Watercress	4	1.5 in. (3.8 cm)	• YYJ-24, 13F-11R, ≠ • Brush: halfway

Direct Seeding Legend

Jang Seeder
X, F, G, Y, YYJ: seed rollers for hoppers
≠: no felt =: with felt
Brush position: Open Halfway Closed

Appendix 2: List of Tools

Tools	Suppliers	Approximate price (USD)
The Essentials		
Broadfork	Growers & Co.	$240
Broadfork with 16" tines (heavy soil work)	Meadow Creature	$300
Bed Preparation (with marker tubes)	Growers & Co.	$121
Bed Preparation		
BCS Walk-Behind Tractor (853 model)	BCS America	$4,700
Rotary Plow	BCS America	$1,700
BCS Power Harrow, 30"	BCS America	$2,500
Tilther	Dubois Agrinovation/Johnny's Selected Seeds	$550
Berta Flail Mower, 34"	Earth Tools	$2,660
Seeders		
Jang seed rollers (set of 5) X-24, F-24, Y-24, LJ-12, G12	Dubois Agrinovation/Johnny's Selected Seeds	$160
Jang Five-Row Seeder (JP-5)	Dubois Agrinovation/Johnny's Selected Seeds	$1,784
Jang Single-Row Seeder (JP-1)	Johnny's Selected Seeds/Dubois Agrinovation	$525
Six-Row Seeder (2nd Edition)	Dubois Agrinovation/Johnny's Selected Seeds	$650
Four-Row Pinpoint Seeder (2nd Edition)	Dubois Agrinovation/Johnny's Selected Seeds	$279
Earthway Seeder	Johnny's Selected Seeds	$168

Tools	Suppliers	Approximate price (USD)
Hoes and Weed Control		
Stirrup hoe, 3¼"	Growers & Co.	$73
Stirrup hoe, 5"	Growers & Co.	$73
Stirrup hoe, 7"	Growers & Co.	$73
Wire hoe, with interchangeable heads	Growers & Co.	$97
Glaser Wheel Hoe, with 12" oscillating hoe	Johnny's Selected Seeds	$465
Flex tine weeder, 30"	Farmers Friend/Johnny's Selected Seeds/Two Bad Cats	$310
Collinear hoe, 7"	Growers & Co.	$73
Bio-discs + Terrateck Double Wheel Hoe	Johnny's Selected Seeds/Dubois Agrinovation	$730
Geotextile or woven ground cover	Dubois Agrinovation	Various prices depending on the size
Harvest Tools		
Opinel No. 10 harvest knife (stainless steel)	Growers & Co./Opinel	$29
Quick-cut Greens Harvester	Dubois Agrinovation/Farmers Friend	$685
Multi-purpose fruit and vegetable harvest lugs (1.75 bu.)	Dubois Agrinovation	$21

Tools	Suppliers	Approximate price (USD)
Cleaning & Packaging Vegetables		
Round trip tote (bin with lid)	Uline	$23
Vented harvest container with handles, to clean vegetables	Dubois Agrinovation	$16
Bubbler	DIY recommended	—
Greens spinner	DIY recommended	—
Shelters		
Row cover (Novagryl 19 g) 22' × 330'	Dubois Agrinovation/Farmers Friend	$125
Caterpillar tunnel, 16' × 100' (hoops every 5')	Dubois Agrinovation/Farmers Friend	$3,270
Gothic caterpillar tunnel, 16' × 100' (hoops every 5')	Farmers Friend	$3,730
Flexible wire hoops, galvanized steel, 64" (200 units)	Dubois Agrinovation	$140
High tunnel 25' × 100' (no automation)	Harnois Industries	$12,600
Greenhouse heating system (propane unit heater + ducts + squirrel cage fan + exhaust pipe + propane tank and line	Harnois Industries	Highly variable
Climate control system	Harnois Industries/Orisha automation	Variable
Weather station for climate control	Harnois Industries/Orisha automation	$2,000
Greenhouse structure, 35' × 100' with arches every 5' + 2 side openings + double door	Harnois Industries	$11,140
Horizontal airflow (HAF) fan	Harnois Industries	$120

241

Tools	Suppliers	Approximate price (USD)
Positive pressure ventilation system for greenhouse	Harnois Industries	$1,000
Temperature and humidity sensor	Harnois Industries / Orisha Automation	$222
Perforated 4 mil. convection tubing, for heating 12" × 100' (6 units)	Harnois Industries	$180
Irrigation controller	Dubois Agrinovation / Orisha Automation	Variable
The Nursery		
Precision Vacuum Seeder	Johnny's Selected Seeds	$599
Vacuum seed plate A128	Johnny's Selected Seeds	$165
Vacuum seed plate A72	Johnny's Selected Seeds	$135
Hand Seed Sower	Johnny's Selected Seeds	$4.65
Dramm ColorStorm hose (⅝" × 100 ft)	Nolt's Greenhouse Supplies / Dramm / Dubois Agrinovation	$88
Dramm watering wand (24 in., 30 degrees)	Nolt's Greenhouse Supplies / Dubois Agrinovation	$8
Brass Dramm valve for watering wand	Nolt's Greenhouse Supplies / Dubois Agrinovation	$15
Dramm Water Breaker 1000	Nolt's Greenhouse Supplies / Dubois Agrinovation	$10
Dramm Water Breaker 400	Nolt's Greenhouse Supplies / Dubois Agrinovation	$10
50-cell trays 21" × 11" × 2¼" (100)	Nolt's Greenhouse Supplies / Bootstrap Farmer	$486
128-cell trays 21" × 11" × 2¼" (100)	Nolt's Greenhouse Supplies / Bootstrap Farmer	$486
72-cell trays 21" × 11" × 2¼" (100)	Nolt's Greenhouse Supplies / Bootstrap Farmer	$486

Tools	Suppliers	Approximate price (USD)
Irrigation		
Irrigation set for farm plot (line, connectors, camlock fittings, hoses, valves)	Dubois Agrinovation / Farmers Friend	$222
Drip tape, 5–6 mil	Dubois Agrinovation	$280
Green flexible lay flat hose 2" × 200' (water transfer)	Dubois Agrinovation	$200
½" Xcel Wobbler kit, Senninger, for 6 × 100' beds (4×)	Dubois Agrinovation/Farmers Friend	$133
Fast-N-Fast Dan sprinkler kit, for 4 × 100' beds (20×)	Dubois Agrinovation/Farmers Friend	$152
Plant Protection		
Insect netting 25 g, (21' × 330')	Dubois Agrinovation	$977
Chemical resistant latex gloves (12 pairs)	Uline	$13
Safety goggles	Uline / 3M / Amazon	$16
3m 5000 P95 respirators + refill kit	Uline	$73
Tyvek® protective suit with hood	Uline	$13
Birchmeier 15-L backpack sprayer	Rittenhouse	$455
Production		
Infinite Dibbler frame (row marker) 32"	Dubois Agrinovation / Johnny's Selected Seeds / Two Bad Cats	$179

Appendix 3: Starting a Fall Garden at Home

For a family or a group (three to five people), a 16' × 40' (5 m × 12 m; 645 ft² [60 m²]) garden can provide enough vegetables to last the entire winter. Because space is limited, each permanent bed should be used for the entire growing season, from May to November. To maximize production (weather permitting), plan to sow a quick-growing crop before each storage crop. In this type of home garden, carrots and potatoes will become the foundations of a winter diet.

If you have the required space and infrastructure, you can start your seedlings indoors, following our recommendations outlined for biointensive market gardeners. If you use this method, remember to include the time spent in trays when determining days to maturity. You can also start everything outdoors through direct seeding. If you choose this option, make sure your vegetables will have enough time to reach maturity before the first frosts hit in your region.

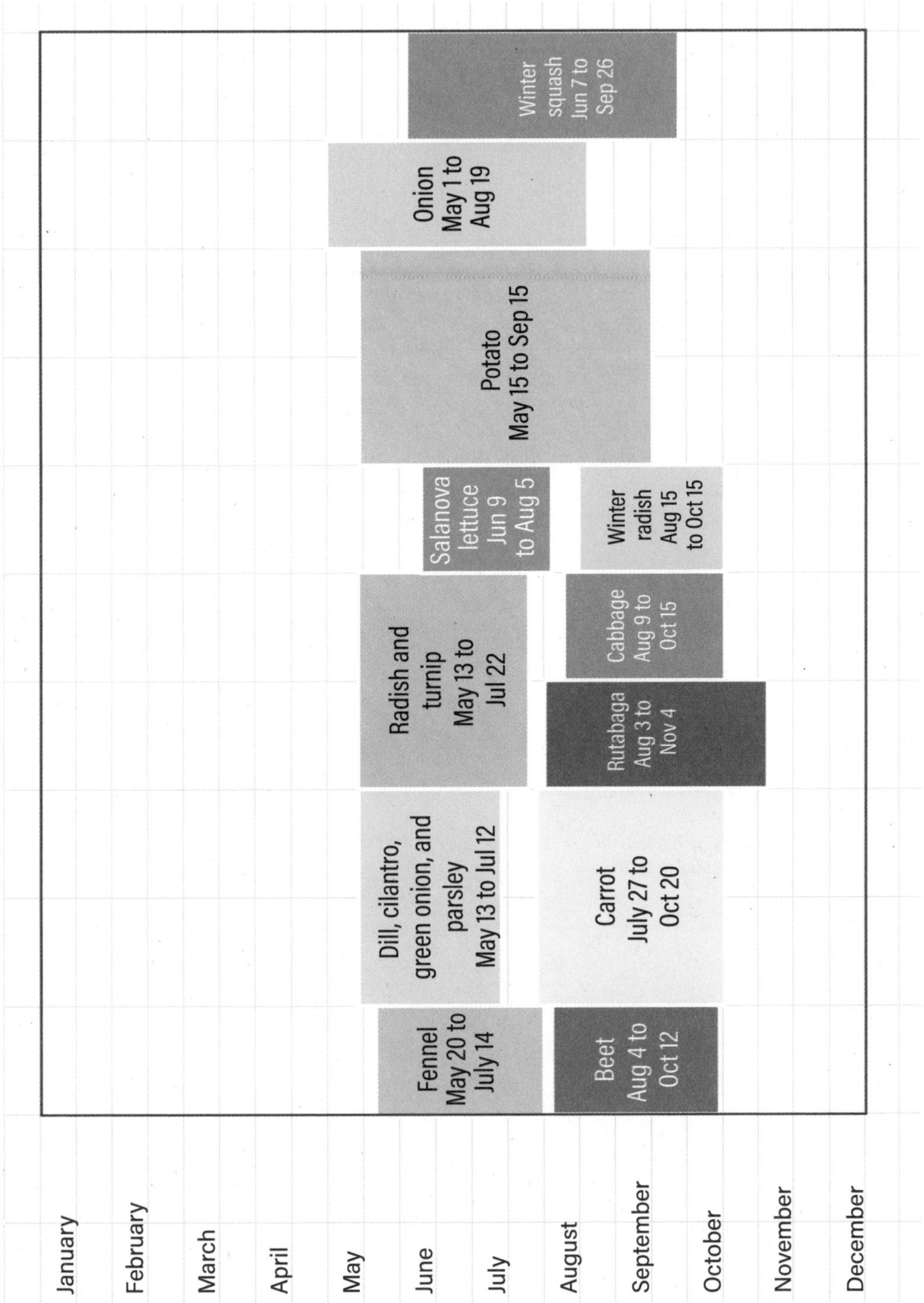

January

February

March

April

May

June

July

August

September

October

November

December

Fennel
May 20 to
July 14

Dill, cilantro,
green onion, and
parsley
May 13 to Jul 12

Radish and
turnip
May 13 to
Jul 22

Salanova
lettuce
Jun 9
to Aug 5

Potato
May 15 to Sep 15

Onion
May 1 to
Aug 19

Winter
squash
Jun 7 to
Sep 26

Winter
radish
Aug 15
to Oct 15

Cabbage
Aug 9 to
Oct 15

Rutabaga
Aug 3 to
Nov 4

Beet
Aug 4 to
Oct 12

Carrot
July 27 to
Oct 20

Appendix 4: Vegetable Classification According to Cold Hardiness

The following table shows the temperatures at which plants will suffer permanent damage. Crops are sorted into three categories according to cold hardiness.

Method	Temperatures (inside the greenhouse) that cause crop damage	Crop
Minimal heating is required	32°F to 25°F (0°C to –4°C)	Arugula Celery New potato
Minimal heating is required	23°F to 14°F (–5°C to –10°C)	Bok choy Chinese cabbage Cilantro Dandelion Green onion Kale (mature) Lettuce Parsley Radish Sorrel Swiss chard (mature) Turnip Watercress
No heating is required (use a row cover at all times when day length is less than 10 hours)	14°F to –22°F (–10°C to –30°C)	Asian greens Baby kale Baby Swiss chard Claytonia Early carrot Komatsuna Mâche Mustard Senposai Spinach

Appendix 5: Starting Spinach Seedling in the Summer

In our experience, starting spinach seedlings in the summer for fall plantings can be quite tricky because the seeds germinate best at low temperatures (under 68°F [20°C]). Seedlings in the summer have low germination rates and can go to seed quite quickly and thus be ruined.

To work around this problem, we have developed the following foolproof method.

- Soak the seeds in water for 24 hours. Place them in a net to drain and dry easily.
- Drain and dry the seeds for 24 hours on a clean towel in the nursery.
- Seed the spinach in 128-cell trays and irrigate.
- Place the trays on racks in a dark and temperate room (≤ 68°F [20°C])
- Wait for seed germination. As soon as 50 percent of the seeds are germinated, move the trays to the coolest zone in your nursery.
- Install a shade cloth over the trays and leave it on until ready to transplant. Transplant seedlings as soon as their roots fill each cell.

Glossary

Aisle
The space between two permanent beds that serves as a walkway. In our fields, this has a width of 18 inches (45 cm), while in the greenhouse it is 12 inches (30 cm). When tending to crops, growers only walk in the aisles to avoid compacting the soil in permanent beds.

Amendment
Substance added to the soil to improve its physical, chemical, and biological characteristics, such as compost, manure, ramial chipped wood (RCW), and lime. Amendments, unlike fertilizers, are used to provide nutrients required for plant growth.

Baby greens
Young greens. To grow baby greens, we densely sow crops like kale and mustard, then harvest them before they have reached maturity, when leaves are still small (length roughly 4 inches [10 cm]).

Baskets
A way for growers to deliver diversified vegetables, weekly, to people who have signed up to be a part of community-supported agriculture (CSA). Throughout the year, members receive baskets that represent their share of the crops, which they purchased at the beginning of the farming season.

Biointensive (agriculture)
An approach that relies on dense plant spacing and prioritizes soil health and biodiversity, to maximize vegetable production in a small area.

Broadfork
U-shaped fork consisting of two wooden handles that support a horizontal bar with many metal tines. This tool relies on the user's body weight to drive the tines deep into the ground. It works the soil and aerates it, without turning it over. When preparing a bed for a new seeding or transplant, this is the first step.

Bubbler
Large tank of water used to wash greens like mesclun, arugula, and spinach. When activated, the bubbling system stirs the greens in the water. This is an efficient way to clean and mix the leaves without too much effort. Next, the leaves are dried and bagged.

Caterpillar tunnel
Temporary shelter made up of a metal structure covered with a transparent plastic film. This shelter is easy to set up and relocate. A single caterpillar tunnel will cover four permanent beds and is perfect for extending the field season. The plastic must be removed from the structure before any major snowfall as the tunnel could collapse under the weight of the snow.

Climacteric (fruit or vegetable)
Adjective describing fruits and vegetables that emit ethylene gas, which accelerates the ripening process of surrounding fruits and vegetables. When planning a fall garden, it's important to consider the presence of ethylene in your cold storage space, where large quantities of vegetables will be kept over several months.

Community-supported agriculture (CSA)

A marketing model in which consumers purchase a share of a farm's production before the start of the agricultural season. The consumer's share is delivered every week, as a vegetable basket. This approach to marketing and sales provides more stable revenue for agricultural operations and builds ties between farmers and their communities.

Cotyledon stage

Stage of plant growth that follows seed germination; this is when cotyledons unfold. Cotyledons are small leaves held within the plant seed. In this stage, the plant is very small and vulnerable. When weeds emerging near a crop are in this stage, it is the perfect time to cultivate.

Cover crop

Crop grown for its positive effects on soil quality and the farm's ecosystem. It is not meant to be harvested, nor sold. Reasons for growing cover crops include covering the soil and preventing erosion, fixing nitrogen (legumes), improving soil structure, and increasing soil organic matter. Once the cover crop is ready, we shred it with a flail mower and cover the plot with a silage tarp. Two to three weeks later, the cover crop residue will have broken down, leaving beds that are ready for the next crop.

Crop plan

A set of tables and charts that is developed in November and December for summer production and in May for winter production to guide crop management over the coming season. The planning process must address the following: developing a planting schedule, creating a calendar for the nursery and direct seedings, mapping out the farm, preparing a seed order, etc.

Crop succession

Production method that is specific to biointensive agriculture, where several crops are grown in the same permanent bed during a growing season or year. For instance, a quick crop (lettuce) is grown and harvested before planting a slow-growing autumn crop (winter squash) in the same bed. The key to success is allowing a two-to-three-week gap between crops. During this time, residue from the first crop is covered with a silage tarp to speed up decomposition.

Cultivar

A plant variety that has been produced through selective breeding, based on yields, specific disease resistance, color, flavor, etc. For northern winter production, choosing the right cultivar is imperative because many have been bred for cold hardiness. When choosing crops, these cultivars should be prioritized.

Days to maturity (DTM)

The number of days expected between a crop's direct seeding or transplant date and the first harvest. To make a crop plan, you will need to know this number. In the fall, DTMs increase exponentially with each passing day, which is relevant when selecting or delaying a planting date.

Direct seeding

Sowing a crop on a permanent bed with the use of a seeder or through broadcast seeding. The crop germinates where it lies and will grow in the bed until it is harvested.

Drainage

This refers to the degree of water retention in any given soil. Well-drained soils retain the water needed for a crop, allowing the rest to flow away. Poorly drained soils retain an excessive amount of water that negatively impacts plant growth. Soil texture and structure will significantly affect drainage. For instance, sand drains quickly and retains little water, while clay is slow to drain and will hold a great deal of water.

Drip irrigation
Irrigation system consisting of thin-walled plastic tubing laid out on the ground. These perforated tubes deliver water right to the base of a plant in a slow and controlled manner. It is the ideal solution for crops in which foliage should be kept dry, which applies to most winter crops.

Early vegetables
Crops harvested in early spring or outside their normal harvest season. By bringing early vegetables to our farmers' markets, we gain a competitive edge. This is why it's necessary to use a range of tools to produce early harvests (i.e., row covers, tunnels, greenhouses, etc.).

Ethylene
This gas speeds up the ripening process in climacteric fruits and vegetables and can cause non-climacteric fruits and vegetables to deteriorate faster. In a cold storage space, ethylene must be monitored; good air circulation helps to prevent any buildup.

Frost-prone (vegetable)
This refers to vegetables that are not frost resistant, which means that they will suffer cell damage in below-freezing temperatures. Tomato plants are an example of a frost-prone vegetable.

Garden
Plot in which crops are grown. In a biointensive garden, a plot usually contains ten permanent beds. When planning crop rotations, it helps to have delineated growing areas.

Geotextile
Water-permeable woven plastic ground cover held in place with metal staples. Holes are made in the fabric for transplanting seedlings according to the required crop spacing. Geotextile significantly limits weeding requirements, which reduces the hours of labor needed to maintain a crop.

Greenhouse
Heated and well-insulated shelter consisting of a metal structure covered with a transparent plastic film. Equipped with a climate control system, the greenhouse is the most complete shelter for winter growing. Most small farms have a greenhouse roughly 100' × 30' (30 m × 9 m).

Ground cover
Material used to cover the ground. Many types of ground covers (straw, geotextile, ramial chipped wood, plastic, etc.) can be used for different objectives (heating or cooling the soil, weed control, amending the soil, etc.).

Hardening off
Sometimes referred to as "acclimation," this process gradually exposes seedlings to harsher conditions, like cold temperatures. It allows plants to become stronger and eventually withstand and survive environmental stresses.

Hardpan
Compacted soil layer generated by repeated tilling. It is typically located just below the tiller's working depth. Hardpans stop water, air, and roots from penetrating deeper soil layers, which can create unfavorable conditions for plant growth.

High tunnel
Simple permanent shelter supporting a transparent plastic film. Unlike greenhouses, high tunnels are not heated and do not contain any climate control systems. High tunnels feature permanent anchors and are not designed to be relocated. Unlike caterpillar tunnels, high tunnels can withstand snow loading, but snow removal may still be required in the event of a big storm.

Insect netting
Production equipment that protects crops from insect pest damage. Insect netting is placed over crops that are vulnerable to certain insects and held in place with rock-filled bags. By using these covers, growers can avoid applying bio-pesticides to crops.

Irrigation
The act of delivering water to a garden using a watering system. In market gardening, crops are irrigated by drip systems or sprinklers.

Low tunnel
Metal structure covered with transparent plastic. This is a very simple low-standing shelter. Easy to install and inexpensive, it is perfect for northern climates, where it can be used to extend the field season in the fall and spring.

Mineralization
This is an essential process for sustaining plant life; the soil organic matter is broken down by microorganisms and macroorganisms releasing nutrients (nitrogen, phosphorus, potassium, etc.) in their mineral form. These nutrients then become available for plant uptake.

Minimal heating
Winter growing method in which heat is used to keep temperatures between 37°F and 41 F (3 C and 5 C) in a greenhouse for cold-hardy vegetables. This method provides a safety net (no vegetables will be lost after a below-freezing night) and eliminates the need for row covers that protect crops from freezing.

Multi-span greenhouse
Sometimes referred to as a multi-chapel greenhouse, this shelter consists of several individual greenhouses installed close together, where the inner walls have been removed. This spacious greenhouse has several V-shaped roofs. The zone below the apex of each roof is called a chapel.

Nursery
Space (usually a greenhouse or a closed room) in which seedlings are sown and grown before being transplanted into permanent beds outside the nursery. To provide the best conditions for seed germination and plant growth, the climate in this space is controlled: 75 F (24 C) during the day and 64°F (18°C) at night. In the fall, we start many winter crops in the nursery, so that summer crops can stay in our greenhouses for as long as possible before being replaced by winter crops.

Occultation
The process of covering beds with a black plastic tarp (silage tarp) to block sunlight. This creates an optimal environment (dark and moist) for soil biological activity, thus speeding up decomposition of residue from vegetable crops and cover crops.

Organic matter
Animal and plant matter located in the soil, which comes in many forms: a bed of living plants, crop residues, a cover crop that is breaking down, living or dead organisms in the soil (earthworms, wood louse, etc.), and humus. All soils have their own measurable organic matter levels, often between 1 and 10 percent.

Overwintering
This is when crops are planted in the fall to be harvested only in the spring. Thanks to a strong root system established in the fall, the crop will be ready for an early spring harvest, as sunlight hours and heat increase. Typically, overwintering occurs in shelters like high tunnels.

Perforated tubing
Tubes that are part of the heating system in a greenhouse. They are placed in the aisles and used to diffuse heat and/or ventilate the crops. To use them, install a blower, equipped with a squirrel cage that sends air into the tubes.

Pest (insect or animal)
Insect or animal that has a detrimental impact on the profitability of agricultural production. A plant protection plan generally needs to be established to mitigate their effects on crops.

Permanent bed
Raised bed with an aisle or walkway on each side. In the field, our beds are 30 inches (75 cm) wide, with aisles spanning 18 inches (45 cm). In our greenhouses, beds are 30 inches (75 cm) wide, with aisles spanning 12 inches (30 cm). By growing crops in beds, we can split the farm into plots containing ten beds each. In winter shelters, permanent raised beds allow the soil to warm up more easily and improve drainage.

Photoperiod
The day length within a 24-hour period. For some vegetables, photoperiod will affect plant growth and maturation. Only certain types of onions, for instance, will produce bulbs in northern countries. This includes long-day (day length is at least 14 hours) or intermediate-day (day length is roughly 12–14 hours) cultivars.

Planting
The step in which the grower sows (direct seeding) or transplants a crop. "Planting a crop" therefore refers to the process of getting a crop started in a bed.

Plant protection
Refers to preventive and corrective strategies established to protect crops from pests and diseases. These strategies take into account a crop's break-even point, to determine whether any action is required. Examples include insect netting, scouting, other physical means of pest control, and—as a last resort—spraying a biopesticide.

Power harrow
BCS walk-behind tractor accessory used to prepare beds for seeding or transplanting. Moving like an eggbeater, the blades work the surface of the soil without turning over any layers.

Precision seeder
Production equipment used to sow crops in a permanent bed. For biointensive agriculture, several models of precision seeders are available, including the Six-Row Seeder and Jang Seeder. These tools are designed to be pushed down the length of the bed, which drives the wheels and releases seeds onto the soil surface with a precise spacing.

Radiant (heating system)
A heating system installed in the ground and made up of pipes containing an antifreeze like glycol. The system heats the liquid, which increases the soil temperature as it runs through the pipes. Radiant heating systems can be used to warm winter crops in a greenhouse.

Rototiller
Accessory that can be mounted onto a BCS walk-behind tractor and used to prepare beds for seeding or transplanting. Blades work deep into the ground, turning over soil layers. When used repeatedly, this tool will degrade the soil structure and, in the long run, generate compaction. The rototiller is suited to shaping soils that are heterogeneous and hard to work.

Row cover
Thermal fabric used to cover crops, protecting them from the cold. Row covers are available in various thicknesses, with thicker materials allowing less light penetration. The P19 cover (0.55 oz./yd.^{2} [19 g/m^2]) is ideal because it blocks only 15 percent of sunlight while still providing adequate insulation in cold weather.

Season extension

An approach to production that aims to lengthen the growing season, despite adverse weather conditions. In a northern climate, the field season can be extended in the fall and spring by a few weeks using simple shelters like row covers, caterpillar tunnels, and low tunnels that protect crops from the wind and cold.

Seeding tray

Production equipment used to start seedlings in a nursery. Trays are referred to according to their number of cells. For winter growing, the most common have 128, 72, and 50 cells; the higher the number is, the smaller the cell. In our nursery, we save the trays with larger cells for winter crops in order to grow bigger seedlings.

Seedling

Small plant started in a nursery for transplanting later into the field or greenhouse. With this approach, we can grow seedlings under conditions that are optimal for seed germination, and we reduce the number of days that a crop will need to spend in the ground.

Silage tarp

Broad black plastic (polyethylene) sheet that is laid out over permanent beds to block sunlight from reaching the soil. The silage tarp has several uses: speeding up the decomposition of crop residues or cover crops, preventing soil erosion, keeping weeds from growing between crop successions, etc.

Soil horizon

A term used in agronomy to refer to a soil layer. All soil is made up of several horizons that, when combined, make up the soil profile. Horizons found deeper in the soil are less fertile and contain less organic matter. Most vegetable roots thus grow within the top layers. When heavy machinery mixes these layers, it has a negative impact on the soil ecosystem.

Sprinkler

Irrigation system consisting of pipes installed perpendicular to the ground. At the top of these pipes, a nozzle sprays water over the crops. This overhead irrigation is easy to install and ideal for watering young transplants or recently seeded beds. With a line of sprinklers, we can simultaneously water eight beds (four beds on each side of the line).

Thermal mass

Material that can store heat and then gradually release it. In warm weather, when the air is hot in the day and cooler at night, the ground can conversely have a cooling effect. This phenomenon is also called "thermal inertia," and it can reduce temperature fluctuations in a greenhouse.

Transplanting

The act of planting seedlings that were started in trays. This method reduces the time that crops will spend in the ground, which maximizes the use of bed space. For winter growing, seedlings are started in the nursery and then transplanted. This allows summer crops to stay in greenhouses for as long as possible before they are replaced by winter seedlings.

True leaf stage

Stage of plant growth that follows the cotyledon stage. At this point, the plant is still small, but it is already becoming established, growing two leaves above the cotyledons. If weeds threatening a crop reach this stage, it means we have fallen behind on weeding, but with the right tools, we can still cultivate the bed. At this critical stage of growth, neglecting to weed will result in a lengthy hand-weeding session.

White-thread stage

Stage of plant growth that follows seed germination and precedes the cotyledon stage. At this time, the plant is quite small, and the stem looks like a white thread. When weeds emerge near a crop and are in this stage, it is the perfect time to cultivate.

253

Bibliography

The Knowledge and Skills of Northern Growers

Gressent, V. A. *Le potager moderne: traité complet de la culture des légumes. Paris*, 1863.

Morneau, J.-G., and J.-J. Daverne. *Manuel pratique de la culture maraîchère de Paris*. Paris, 1845.

Vercier, J. *Culture Potagère*. Paris: Hachette, 1911.

Protecting Crops with Simple Shelters

Ackerman, S. A., and J. Knox. *Meteorology: Understanding the Atmosphere*. John & Bartlett Learning, 2015.

Kyba, C. C. M., et al. "Artificially Lit Surface of Earth at Night Increasing in Radiance and Extent." *Science Advances*, Vol. 3, no. 11 (November 2017). Also available online: https:// advances. sciencemag.org/content/3/11 /e1701528

Organic Federation of Canada. "Artificial Light as a Supply to Natural Light." Organic Federation of Canada: 2020. https://organicfederation .ca/resource/final-questions-answers-canadian -organic-standards/

Tending to Vegetables in Winter

Adams, J. B. "Aphid Survival at Low Temperatures," *Canadian Journal of Zoology*, Vol. 40, no. 6 (1962): 951–956.

Amare, G., and B. Desta. "Coloured Plastic Mulches: Impact on Soil Properties and Crop Productivity," *Chemical and Biological Technologies in Agriculture*, Vol. 8, no. 4 (2021).

"Blanc (syn. Oïdium) – Laitues." IRIIS Phyto- protection Online (French only): www.iriisphyto protection.qc.ca/Fiche/Champignon?image Id=9000

Denlinger, D. L., and R. E. Lee, eds. *Low Temperature Biology of Insects*. Cambridge University Press, 2010.

Gheshm, R., and R. Nelson Brown. "The Effects of Black and White Plastic Mulch on Soil Temperature and Yield of Crisphead Lettuce in Southern New England." *HortTechnology*, Vol. 30, no. 6 (2020): 781–88.

Grzyb, A., A. Wolna-Maruwka, and A. Niewiad- omska. "Environmental Factors Affecting the Mineralization of Crop Residues," *Agronomy*, Vol. 10, no. 12 (2020): 1951.

Guntiñas, M. E., M. C. Leirós, C. Trasar-Cepeda, and F. Gil-Sotres. "Effects of Moisture and Tem- perature on Net Soil Nitrogen Mineralization: A Laboratory Study," *European Journal of Soil Biology*, Vol. 48 (2012): 73–80.

Labrie, G., J. E. Maisonhaut, and L. Lambert. "Affiches des auxiliaires de lutte en serre (Aleurodes, Pucerons, Mouches noires, Thrips, Tétranyques)." Centre de recherche en agroalimentaire de Mirabel, 2020.

"Mildiou - Laitues." Iriis Phytoprotection. Online (French only): www.iriisphytoprotection.qc.ca /Fiche/Champignon?imageId=9011

Miura, M. "Effect of Climate Variability and Extreme Events on Microbial Activity in Soils." PhD diss., Bangor University, 2020.

Orzolek, M., Ed. *A Guide to the Manufacture, Performance, and Potential of Plastics in Agri- culture*. Elsevier Science & Technology Books, 2017.

Pietikäinen, J., M. Pettersson, and E. Bååth. "Comparison of Temperature Effects on Soil

Respiration and Bacterial and Fungal Growth Rates," *FEMS Microbiology Ecology*, Vol. 52, no. 1 (2005): 49–58.

Réseau d'avertissement phytosanitaire. "Le nettoyage et la désinfection des serres en fin de saison." *Cultures en serres*, no. 14 (September 2012). Online (French only): www.agrireseau .net/Rap/documents/b14cs12.pdf

Snyder, K., A. Grant, C. Murray, and B. Wolff. "The effects of Plastic Mulch Systems on Soil Temperature and Moisture in Central Ontario," *HortTechnology*, Vol. 25, no. 2 (2015): 162–170.

Tougeron, K., et al. "Effect of Diapause on Cold-Resistance in Different Life-Stages of an Aphid Parasitoid Wasp." Plant Biology Research Institute, Department of Biological Sciences, Université de Montréal, 2018.

Planning a Winter Garden
"Caveau à legumes: Répertoire du patrimoine culturel du Quebec." Department of Culture and Communications of Quebec. Online (French only): www. patrimoine-culturel.gouv .qc.ca/rpcq/detail.do?methode=consulter&id= 178203&type=bien

Food Safety Network. *Safe Preparation and Storage of Aboriginal Traditional/Country Foods: A Review*. National Collaborating Center for Environmental Health. https://ncceh.ca/sites /default/files/Aboriginal_Foods_Mar_2009.pdf

Johnny's Selected Seeds. "Post-Harvest Handling & Storage Guidelines for Storage Crops." www.johnnyseeds.com/growers-library /vegetables/storage-crops.html

La France, D. *La culture biologique des legumes*. Berger, 2010.

"Les caveaux à légumes, Fiches techniques sur les systèmes alimentaires de proximité" Équipements et infrastructures alimentaires, Agri-Réseau. Online (French only): www.agri reseau.net/documents/Document_98614.pdf

Growing Handbooks—Useful Resources
In developing our winter farming methods, we drew on both old and new resources. Each one is sure to be a valuable reference for anyone who might have an interest in this topic.

Carnavalet, C. *Agriculture biologique, une approche scientifique*. Paris: Éditions France Agricole, 2018.

Coleman, E. *The Winter Harvest Handbook: Year-Round Vegetable Production Using Deep-Organic Techniques and Unheated Greenhouses*. Chelsea Green, 2009.

De Carné-Carnavalet, C. *Biologie du sol et agriculture durable: une approche organique et agroécologique*, Paris: Éditions France Agricole, 2015.

École supérieure d'agriculture. *Manuel d'agriculture par les professeurs de l'École supérieure d'agriculture de Ste-Anne de La Pocatière: Les champs*. Ateliers de l'Action catholique, 1947.

Frost, J. *The Living Soil Handbook: The No-till Grower's Guide to Ecological Market Gardening*. Chelsea Green, 2021.

Gressent, V. A. *Le potager moderne: traité complet de la culture des légumes*. Paris, 1863.

Institut de l'agriculture et de l'alimentation biologiques. *Produire des légumes biologiques tome 1: Généralités et principes techniques*, 2017; *Produire des légumes biologiques tome 2: Fiches techniques par légume*. 2017.

Morneau, J.-G., and J.-J. Daverne. *Manuel pratique de la culture maraîchère de Paris*. Paris, 1845.

Paillieux, A. *Le potager d'un curieux: histoire, culture et usages de 100 plantes comestibles peu connues ou inconnues*. Paris: Librairie agricole de la maison rustique, 1885.

Vercier, J. *Culture Potagère*. Paris: Hachette, 1911.

255

Index

About the Authors

JEAN-MARTIN FORTIER is a farmer, educator, and advocate for regenerative agriculture. He is author of the international bestseller *The Market Gardener*, founder of Growers & Co., and co-founder of the Market Gardener Institute. In 2015 he established the research farm la Ferme des Quatre-Temps. He lives and farms in Quebec, Canada.

CATHERINE SYLVESTRE is a professional agronomist and director of vegetable production and leader of the market garden team at la Ferme des Quatre-Temps. She develops, implements, and teaches best practices for cold season growing, specializing in crop protection and greenhouse production for northern climates. She lives in Quebec, Canada.

ABOUT NEW SOCIETY PUBLISHERS

New Society Publishers is an activist, solutions-oriented publisher focused on publishing books to build a more just and sustainable future. Our books offer tips, tools, and insights from leading experts in a wide range of areas.

We're proud to hold to the highest environmental and social standards of any publisher in North America. When you buy New Society books, you are part of the solution!

At New Society Publishers, we care deeply about *what* we publish—but also about *how* we do business.

• This book is printed on 100% **post-consumer recycled paper**, processed chlorine-free, with low-VOC vegetable-based inks (since 2002).

• Our corporate structure is an innovative employee shareholder agreement, so we're one-third employee-owned (since 2015)

• We've created a Statement of Ethics (2021). The intent of this Statement is to act as a framework to guide our actions and facilitate feedback for continuous improvement of our work

• We're carbon-neutral (since 2006)

• We're certified as a B Corporation (since 2016)

• We're Signatories to the UN's Sustainable Development Goals (SDG) Publishers Compact (2020–2030, the Decade of Action)

To download our full catalog, sign up for our quarterly newsletter, and to learn more about New Society Publishers, please visit newsociety.com

ENVIRONMENTAL BENEFITS STATEMENT

New Society Publishers saved the following resources by printing the pages of this book on chlorine free paper made with 100% post-consumer waste.

TREES	WATER	ENERGY	SOLID WASTE	GREENHOUSE GASES
121	9,300	51	420	52,500
FULLY GROWN	GALLONS	MILLION BTUs	POUNDS	POUNDS

Environmental impact estimates were made using the Environmental Paper Network Paper Calculator 4.0. For more information visit www.papercalculator.org

Certified
B Corporation

new society
PUBLISHERS
www.newsociety.com

FSC
www.fsc.org
MIX
Paper from responsible sources
FSC® C016245

SDG PUBLISHERS COMPACT